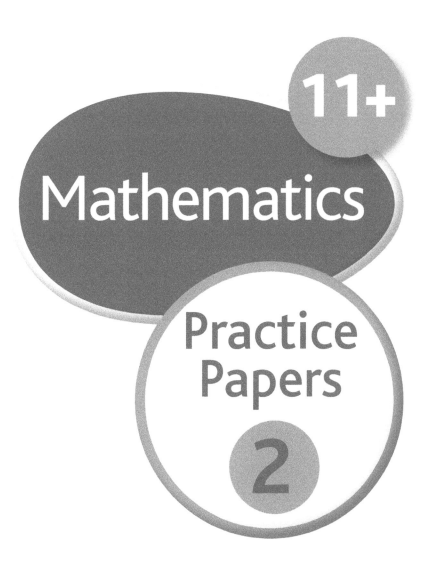

# 11+

# Mathematics

## Practice Papers

## 2

Hachette UK's policy is to use papers that are natural, renewable and recyclable products and made from wood grown in well-managed forests and other controlled sources. The logging and manufacturing processes are expected to conform to the environmental regulations of the country of origin.

Orders: **Teachers** Please contact Hachette UK Distribution, Hely Hutchinson Centre, Milton Road, Didcot, Oxfordshire, OX11 7HH. Telephone: (44) 01235 400555. Email primary@hachette.co.uk. Lines are open from 9 a.m. to 5 p.m., Monday to Friday.
**Parents, Tutors** please call: 020 3122 6405 (Monday to Friday, 9:30 a.m. to 4.30 p.m.).
Email: parentenquiries@galorepark.co.uk
Visit our website at www.galorepark.co.uk for details of other revision guides for Common Entrance, examination papers and Galore Park publications.

**ISBN: 978 1 471869 05 1**

© Steve Hobbs 2016
First published in 2016 by
Galore Park Publishing Ltd
An Hachette UK Company
Carmelite House
50 Victoria Embankment
London EC4Y 0DZ
www.galorepark.co.uk
Impression number    10  9  8  7
Year        2025  2024  2023  2022  2021

Typeset in India
Printed in the UK
Illustrations by Integra Software Services, Ltd.

A catalogue record for this title is available from the British Library.

Name: _____

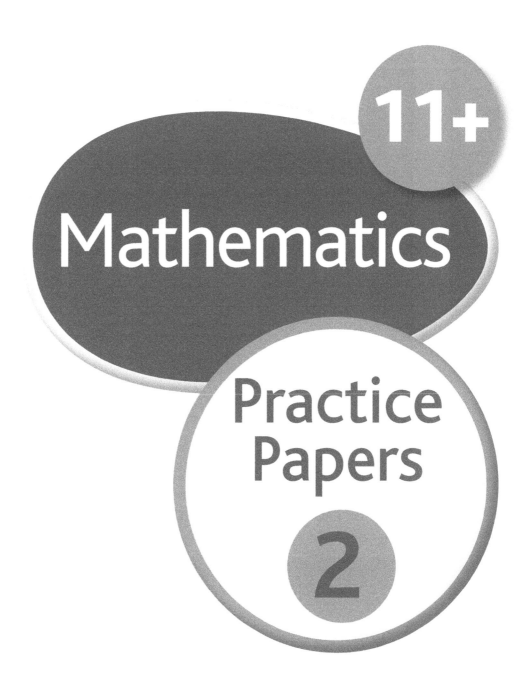

# 11+

# Mathematics

# Practice Papers

# 2

**Steve Hobbs**

GALORE PARK

AN HACHETTE UK COMPANY

# Contents and progress record

*Practice Papers 1* contains papers 1–9 and should be attempted first

*Practice Papers 2* contains papers 10–13 and should be attempted after *Practice Papers 1*

| Speed | Question type | Score | Time |
|---|---|---|---|
| Slow | Multiple-choice | / 29 | : |
| Average | Multiple-choice | / 36 | : |
| Slow | Standard | / 29 | : |
| Average | Standard | / 29 | : |
| Fast | Standard | / 29 | : |
| Fast | Standard | / 67 | : |
| Fast | Standard | / 54 | : |
| Fast | Standard | / 51 | : |
| Average | Multiple-choice | / 96 | : |

| Speed | Question type | Score | Time |
|---|---|---|---|
| Realistic | Standard | / 98 | : |
| Average | Multiple-choice | / 114 | : |
| Realistic | Standard | / 192 | : |
| Fast | Standard | / 174 | : |

# How to use this book

## Introduction

These *Practice Papers* have been written to provide final preparation for your 11+ Maths test.

*Practice Papers 1* includes nine model papers with a total of 184 questions. There are ...

- four training tests, which include some simpler questions and slower timing designed to develop confidence
- four tests in the style of pre-tests, ISEB (Independent Schools Examination Board) and short-format CEM (Centre for Evaluation and Monitoring)/bespoke tests in terms of difficulty, speed and question variation
- one test in the style of the longer format GL (Granada Learning)/bespoke tests in multiple-choice question format.

*Practice Papers 2* includes four model papers with a total of 260 questions. These papers include ...

- one in the style of a longer-format GL bespoke test in standard question format
- three further longer-format tests with challenging content, speed and question variation to support all 11+ tests.

*Practice Papers 2* will help you ...

- become familiar with the way long-format 11+ tests are presented
- build your confidence in answering the variety of questions set
- work with the most challenging questions set
- tackle questions presented in different ways
- build up your speed in answering questions to the timing expected in the most demanding 11+ tests.

## Pre-test and the 11+ entrance exams

The Galore Park 11+ series is designed for pre-tests and 11+ entrance exams for admission into Independent Schools. These exams are often the same as those set by local Grammar Schools too. 11+ Maths tests now appear in different formats and lengths and it is likely that if you are applying for more than one school, you will encounter more than one of type of test. These include:

- pre-tests delivered on-screen
- 11+ entrance exams in different formats from GL, CEM and ISEB
- 11+ entrance exams created specifically for particular independent schools.

Tests are designed to vary from year to year. This means it is very difficult to predict the questions and structure that will come up, making the tests harder to revise for.

To give you the best chance of success in these assessments, Galore Park has worked with 11+ tutors, independent school teachers, test writers and specialist authors to create these *Practice Papers*. These books cover the styles of questions and the areas of Maths that typically occur in this wide range of tests.

*Because 11+ tests now aim to include variations in the content and presentation of questions, making them increasingly difficult to revise for, **both** books should be completed as essential preparation for all 11+ Maths tests.*

## For parents

These *Practice Papers* have been written to help both you and your child prepare for both pre-test and 11+ entrance exams.

For your child to get maximum benefit from these tests, they should complete them in conditions as close as possible to those they will face in the exams, as described in the 'Working through the book' section below.

Timings get shorter as the book progresses to build up speed and confidence.

Some of these timings are very demanding and reviewing the tests again after completing the books (even though your child will have some familiarity with the questions) can be helpful, to demonstrate how their speed has improved through practice.

## For teachers and tutors

This book has been written for teachers and tutors working with children preparing for both pre-test and 11+ entrance exams. The variations in length, format and range of questions is intended to prepare children for the increasingly unpredictable tests encountered, with a range of difficulty developed to prepare them for the most challenging and on-screen adaptable tests.

# Working through the book

The **Contents and progress record** helps you to understand the purpose of each test and track your progress. Always read the notes in this record before beginning a test as this will give you an idea of how challenging the test will be!

You may find some of the questions hard, but don't worry. These tests are designed to build up your skills and speed. Agree with your parents on a good time to take the test and set a timer going. Prepare for each test as if you are actually going to sit your 11+ (see 'Test day tips' on page 8):

- Complete the test with a timer, in a quiet room, noting down how long it takes you, writing your answers in pencil. Even though timings are given, you should complete ALL the questions.
- Mark the test using the answers at the back of the book.
- Go through the test again with a friend or parent and talk about the difficult questions.
- Have another go at the questions you found difficult and read the answers carefully to find out what to look for next time.

The **Answers** are designed to be cut out so that you can mark your papers easily. Do not look at the answers until you have attempted a whole paper. Each answer has a full explanation so you can understand why you might have answered incorrectly and how to award the marks.

When you have finished a test, turn back to the Contents and progress record and fill in the boxes:

- write your total number of marks in the 'Score' box
- note the time you took to complete ALL the questions in the 'Time' box.

After completing both books you may want to go back to the earlier papers and have another go to see how much you have improved!

# Continue your learning journey

When you've completed these *Practice Papers*, you can carry on your learning right up until exam day with the following resources.

 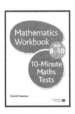 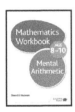 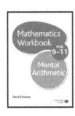    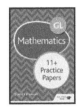

The *Revision Guide* reviews all the areas of Maths you may encounter in your 11+ entrance exams as well as tips and guidance to excel above the other candidates. This includes additional material required by some schools that you may not have encountered in schools.

*Practice Papers 1* contains four training papers, four short-format papers and one longer length paper with answers to improve your accuracy, speed and ability to deal with a wide range of questions under pressure.

*CEM 11+ Mathematics Practice Papers* test and encourage you in preparation for the mathematics content of CEM 11+ tests, including the bespoke tests created by CEM for individual schools. The four papers will help to assess your knowledge, skills, understanding and reasoning ability across all areas of mathematics.

*GL 11+ Mathematics Practice Papers* contains four practice papers designed to prepare you for the GL-style tests. Each paper has 50 questions in line with the actual GL test format.

The two *Workbooks* (*Mental Arithmetic* and *10-Minute Maths Tests*) will further develop your skills with 50 mental arithmetic tests and 80 10-minute tests to work through. These titles include more examples of the different types of questions you meet in these *Practice Papers* – the more times you practise the questions, the better equipped for the exams you will be.

# Paper 10

1 Write these numbers in order of size, starting with the largest. (1)

| 7.07 | 0.71 | 1.7 | 7.1 | 17.1 |

_____ _____ _____ _____ _____

2 Circle all the numbers in the list below that are between 4.4 and 4.7 (2)

$4\frac{3}{7}$    $4\frac{1}{2}$    $4\frac{3}{4}$    $4\frac{2}{3}$    $4\frac{4}{9}$

3 These two years are written in Roman numerals.

**MMXXXV** and **MML**

How many years are there between these two years? _____ (3)

4 Whilst Mike was based at Vostok Station, Antarctica, the temperature ranged from −18 °C to −72 °C.

What is the difference between these two temperatures? _____ (1)

5 Circle all the amounts that can be made with four coins. (1)

68p        87p        92p        £1.18        £1.21

6 The average distance between the moon and the Earth is 238 857 miles.

Round this distance to the nearest 10 000 miles. _____ (1)

7 How many of the numbers below are multiples of 7? _____ (1)

28        44        84        107        112        161

8 Mark, Jed and Rick are running a race. Mark finishes in 145 seconds. Jed finishes 18 seconds after Mark. Rick finishes 11 seconds before Jed.

What is Rick's finishing time in minutes and seconds? _____ (2)

9 Zoe gets home from school at 3:45 in the afternoon. She gets changed, which takes 5 minutes, and watches television for 20 minutes. She then walks for 15 minutes to reach her gymnastics lesson. The lesson lasts 1 hour. She then walks straight home.

At what time does she get home from her gymnastics club? _____ (2)

10 If I quadruple a number and then add 7, I get the same answer as when I multiply the number by 5

What is the original number? _____ (3)

11 In this pyramid, the expression in each brick is formed by adding the terms in the two bricks below it. Complete this pyramid and write down the expression in the top brick. (3)

| 3a | 2 | e | 5 |

**Turn over to the next page**

12 Jenny is making a quilt out of small pieces of material. The diagram shows the dimensions of the quilt and the small pieces of material. How many small pieces of material will she need to make the quilt? _____ (3)

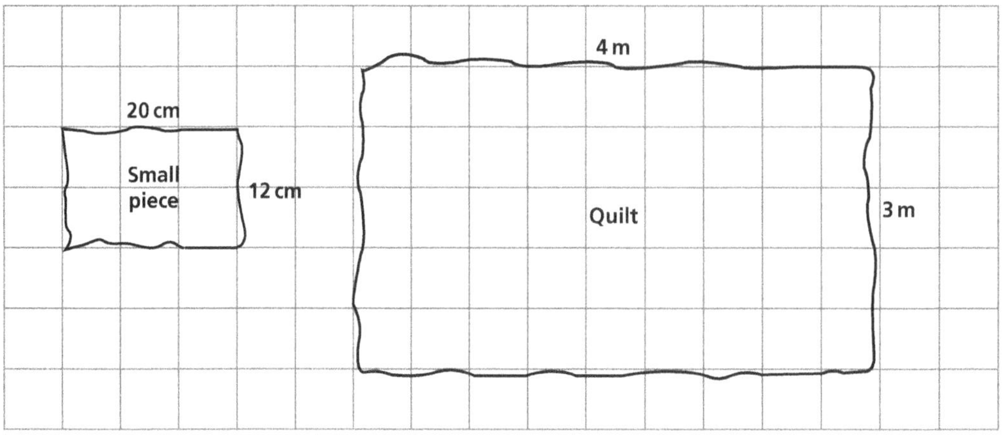

20 cm

Small piece

12 cm

4 m

Quilt

3 m

*Not drawn accurately*

13 Complete the square so that the numbers in each row, column and diagonal add up to 15 Each cell must contain one number between 1 and 9 and each number can be used only once. (2)

| 2 |  | 4 |
|---|---|---|
|   |   |   |
|   | 1 |   |

14 Joshua and Darren share their bag of 28 sweets in the ratio 4:3
How many more sweets does Joshua get than Darren? _____ (2)

15 Jamie is buying a new table. The full price is £670, but it is reduced in the New Year sale by 20%. How much will Jamie pay for the table in the New Year sale?
_____ (3)

16 Kathryn drives a different distance each day for four days. The distances are: 47 km, 58 km, 28 km and 47 km. If she uses 1 litre of fuel for every 15 km she drives, how much fuel does she use each day, on average, over the four days? _____ (3)

17 An adult ticket to watch our local football team costs £$A$. A child ticket is half the price of an adult ticket. What is the total cost for 6 adults and 4 children to watch the local football team? _____ (3)

18 A bag contains 60 red sweets, 40 yellow sweets, 70 blue sweets and 30 orange sweets. I pick a sweet without looking. What is the probability I pick a red sweet?
_____ (2)

19 A packet of felt-tip pens costs £1.15 and a packet of pencils costs 98p. What do 3 packets of felt-tip pens and 2 packets of pencils cost? _____ (3)

20 Write each number from the list below in the correct region of the Carroll diagram. (1)

8          41          27          48          21          24          18          17          15

|                    | Multiple of 4 | Not a multiple of 4 |
|--------------------|---------------|---------------------|
| Not a multiple of 3 |               |                     |
| Multiple of 3       |               |                     |

21 $\frac{1}{2}$ × 136.48 = _____ (1)

10

22 To get from A to B, Ranulph must travel along the line given by the equation $y = 2x - 1$. Plot his route on the co-ordinate grid using values of $x$ from 1 to 5  (1)

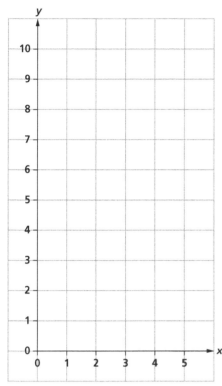

23 The Makkah Royal Clock Tower Hotel in Saudi Arabia has 120 floors. Aamir is on floor 48 How far up the hotel is he? Write your answer as a fraction in its lowest terms.

_____  (1)

24 Circle the probability word from the list below that best describes the likelihood that Wednesday will follow Tuesday.  (1)

**certain   very likely   likely   even chance   unlikely   very unlikely   impossible**

25 How many millilitres of liquid are in this container? _____  (1)

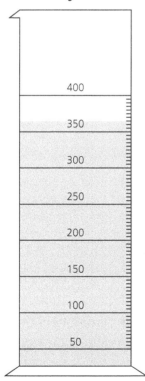

**Turn over to the next page**

26 A 10p coin has a diameter of 2.45 cm. What is the total length of a line of ten 10p coins, as shown in the diagram? _____ (1)

*Not drawn accurately*

27 What is the area of this parallelogram? _____ (1)

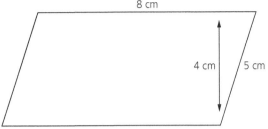

*Not drawn accurately*

28 $870 \div n = 0.87 \times 10$

Work out the value of $n$? _____ (2)

29 Use any appropriate method to multiply 2863 by 6 _____ (1)

30 Divide 8736 by 6 _____ (1)

31 Work out the area of this rectangle. _____ (2)

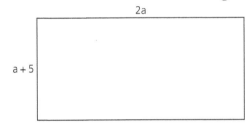

32 Reflect the shape in the mirror line. (1)

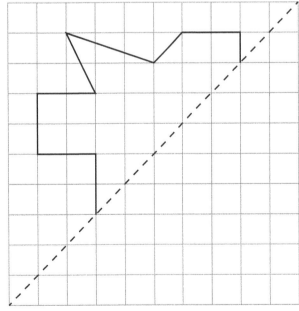

33 Tanya has forgotten the combination for her bicycle lock. Her mum remembers that:

- The number is between 150 and 170
- The number is in the 7 times table
- The digits in the number add up to 8

What is the combination? _____ (3)

**34** Use this information to complete the Venn diagram. (4)
  • There are 32 children in class 4H.
  • 16 children play cricket only.
  • 4 children play cricket and rounders.
  • 3 children don't play either sport.

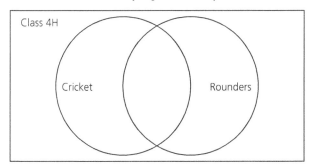

Class 4H

Cricket     Rounders

**35** What is the missing number in the calculation below? _____ (2)
275 + ? + 78 = 568

**36** It takes Hal 1 hour 20 minutes to drive the 32 miles from London Victoria to Crawley. He drives at a constant speed throughout the journey. How far does he drive in 1 hour? _____ (2)

**37** The graph shows the average monthly temperature in Brighton during one year. What is the range of the temperatures? _____ (3)

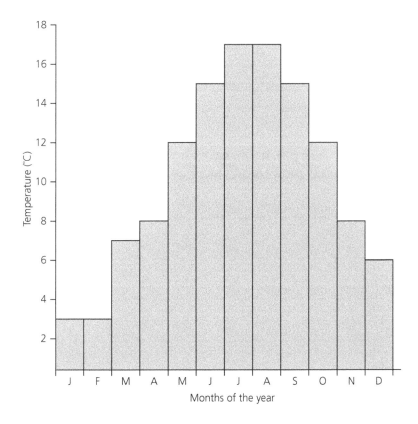

Months of the year

**Turn over to the next page**

38 Work out the size of angle $x$. _____ (2)

*Not drawn accurately*

39 Charlotte visits a café and buys a cake that costs 95p, a cup of tea that costs £1.80 and a chocolate bar that costs 68p. How much is this altogether? _____ (1)

40 Use the grid method to work out 45 × 38

Write your working on the grid below. (1)

41 What is the the size of angle $x$? _____ (1)

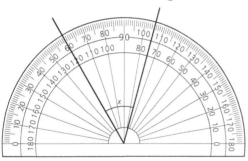

42 Gordon wants to serve his dessert at 7:30 pm. It will take him 2 hours 40 minutes to prepare and cook. At what time should he start making the dessert? _____ (2)

43 The grid contains some letters. Write down, in order, the co-ordinates of the letters that spell ALICE. (5)

    A        L        I        C        E

(__ , __) (__ , __) (__ , __) (__ , __) (__ , __)

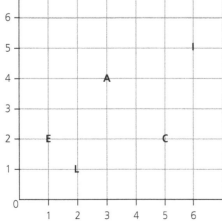

44 What is the chance that when I roll a die, I roll an even number? _____ (2)

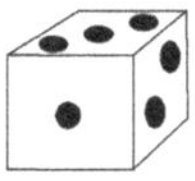

45 Joe walks 3.2 km before lunch and 1826 m after lunch. How many metres does he walk in total? _____ (2)

46 Alfie asked a group of children what their favourite subject was. His results are shown in the pictogram.

Key:  ☺ 2 Childern

                                              ☾ 1 Child

English     ☺  ☺  ☺  ☾

Maths       ☺  ☺  ☺  ☺  ☺  ☺

Science     ☺  ☺  ☾

History     ☺  ☺  ☺

Geography  ☺  ☺  ☺  ☾

How many children were in the group? _____ (2)

47 What does $F$ equal in the equation: $6(F - 7) = 6$

$F =$ _____ (3)

48 What does $W$ equal in the equation: $4(2W + 3) = 4(W + 2)$

$W =$ _____ (4)

49 These are the batting scores for one cricketer in a test series.

**49, 35, 109, 78, 38, 0, 145, 5, 189, 27**

What is his mean score? _____ (2)

50 Edmund is walking up a mountain path. The graph shows his ascent (climb up) and descent (climb down).

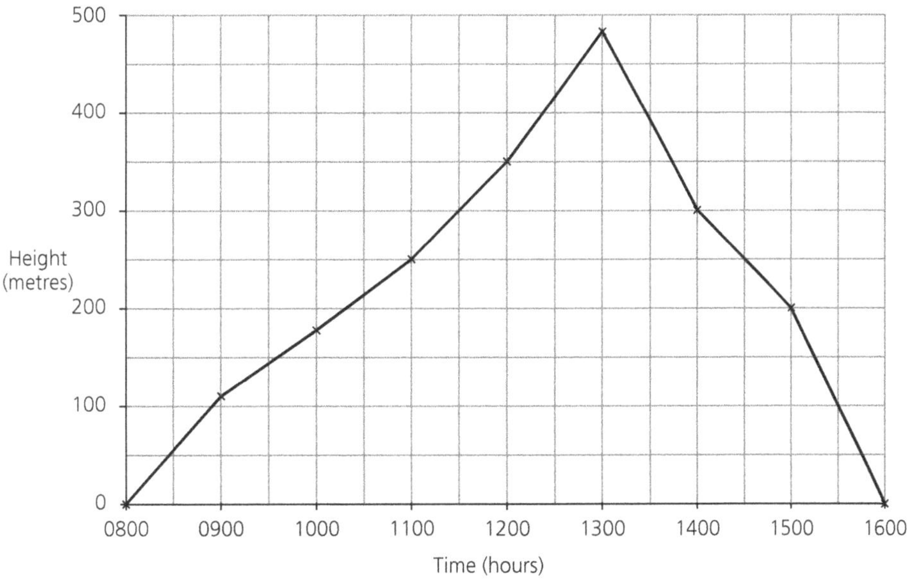

Height (metres) vs Time (hours)

During which hour does Edmund climb the greatest height? _____ (1)

**Record your results and move on to the next paper**

Score ☐ / 98   Time ☐ : ☐

# Paper 11

**Circle the correct answer for each question.**

1 The attendance at a football match at Wembley Stadium was given as 88 000, to the nearest 1000
  The actual attendance figure is given below. Which one is it? (1)
  (a) 87 241     (b) 88 748     (c) 88 500     (d) 88 237     (e) 88 501

2 John trains with his local running club on a 400 m running track. This week John needs to run 10 km. How many laps of the track is this? (2)
  (a) 10     (b) 15     (c) 20     (d) 25     (e) 30

3 An alarm flashes intermittently to show that it is working. It flashes at 08:00 and then at 20-second intervals. How many times will it flash between 08:00 and 10:00? (3)
  (a) 180     (b) 240     (c) 360     (d) 361     (e) 480

4 A farmer keeps pigs and chickens. He has 32 animals altogether and the animals have a total of 80 legs. How many chickens does the farmer have? (5)
  (a) 16     (b) 18     (c) 8     (d) 14     (e) 24

5 Which shape is in the wrong part of the Carroll diagram? (1)
  (a) Square               (b) Rectangle               (c) Equilateral triangle
  (d) Isosceles triangle        (e) Kite

|  | At least one pair of parallel sides | No parallel sides |
|---|---|---|
| All sides same length | Square | Equilateral triangle |
| Not all sides same length | Kite<br>Rectangle | Isosceles triangle |

6 Which one of these events will certainly happen? (1)
  (a) I will get something wrong tomorrow.
  (b) My brother's favourite team will win soon.
  (c) Tuesday will follow Monday.
  (d) The total of two dice will be 13
  (e) It will rain during the summer.

7 What is $\frac{3}{4}$ of 92? (1)
  (a) 76     (b) 69     (c) 72     (d) 78     (e) 63

8 What is 89 × 32? (1)
  (a) 2884     (b) 2848     (c) 2448     (d) 2284     (e) 2484

9 What is 1376 ÷ 16? (1)
  (a) 48     (b) 86     (c) 74     (d) 60     (e) 68

10 What is the equation of the line shown on the graph? (2)

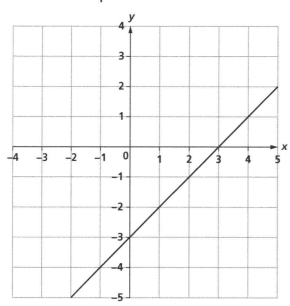

(a) $y = x + 1$    (b) $y = 2x + 1$    (c) $y = x - 2$    (d) $y = x - 3$    (e) $y = x - 1$

11 In a gymnastics club, there are 8 girls for every 3 boys. There are 55 children in the club. How many boys are there? (3)

(a) 15 boys    (b) 40 boys    (c) 11 boys    (d) 20 boys    (e) 24 boys

12 When making blackcurrant squash, Margaret mixes 1 part of the concentrate with 15 parts of water. How much concentrate does she use to make 160 litres of blackcurrant squash? (2)

(a) 10 litres    (b) 15 litres    (c) 18 litres    (d) 20 litres    (e) 21 litres

13 Which percentage is equivalent to the fraction $\frac{170}{250}$ ? (2)

(a) 17%    (b) 34%    (c) 42%    (d) 68%    (e) 85%

14 What do you need to add to 7.67 to make 34.45? (1)

(a) 42.12    (b) 23.48    (c) 27.48    (d) 24.78    (e) 26.78

15 The shape shown is made up of 3 equilateral triangles on each side of a square that has sides of 18 cm. What is the perimeter of the shape? (2)

(a) 144 cm    (b) 72 cm    (c) 24 cm    (d) 108 cm    (e) 100 cm

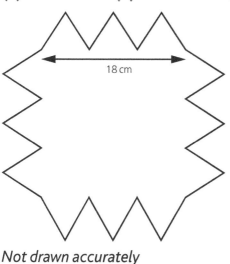

18 cm

*Not drawn accurately*

**Turn over to the next page**

16 Arjun is reading a book. He has read 135 pages, which is $\frac{3}{8}$ of the book. How many pages does the book have altogether? (2)

(a) 450 pages   (b) 360 pages   (c) 320 pages   (d) 400 pages   (e) 380 pages

17 Mellissa is putting a medal on a ribbon for each of the 28 junior club members. Each piece of ribbon is 70 cm long. How many metres of ribbon does she use altogether? (2)

(a) 19.6 m   (b) 196 m   (c) 1.96 m   (d) 0.196 m   (e) 0.0196 m

18 There are between 80 and 90 biscuits in a biscuit tin. Tommy eats exactly $\frac{1}{4}$ of the biscuits and Billy eats exactly $\frac{1}{7}$ of them. How many biscuits were in the tin to start with? (2)

(a) 81   (b) 84   (c) 85   (d) 87   (e) 89

19 What is the size of angle $x$? (2)

(a) 74°   (b) 126°   (c) 54°   (d) 63°   (e) 45°

Not drawn accurately

20 What is the size of angle $y$? (2)

(a) 39°   (b) 61°   (c) 85°   (d) 46°   (e) 44°

Not drawn accurately

21 George is going on holiday to the USA. He wants to take $900 dollars with him.
The exchange rate is £1 buys $1.45
How much will this cost him, to the nearest pound? (2)

(a) £1305   (b) £621   (c) £620   (d) £900   (e) £600

22 What is $\frac{3}{5} + \frac{2}{6}$? (3)

(a) $\frac{4}{5}$   (b) $\frac{5}{6}$   (c) $\frac{14}{30}$   (d) $\frac{14}{15}$   (e) $1\frac{1}{5}$

23 What is $1\frac{3}{4} - \frac{2}{5}$? (2)

(a) $1\frac{1}{4}$   (b) $1\frac{7}{20}$   (c) $1\frac{1}{5}$   (d) $1\frac{1}{10}$   (e) $1\frac{3}{20}$

24 Chairs are being set out in rows of 26 for the school play. There are 403 parents coming. How many rows of chairs are needed? (2)

(a) 16   (b) 15   (c) 26   (d) 20   (e) 14

25 Khalid is running a marathon in Europe and so the distance is measured in kilometres, rather than miles. He gets to the 18 km mark. How far has Khalid run in miles?
(1 mile = 1.6 km) (1)

(a) 28.8 miles   (b) 11.25 miles   (c) 16 miles   (d) 12 miles   (e) 14 miles

26 A burger van outside a football ground sells 158 burgers for £3 each on match day? How much money do they take? (1)

(a) £474 (b) £553 (c) £316 (d) £632 (e) £550.50

27 Which of these is a square number? (1)

(a) 404 (b) 200 (c) 225 (d) 1000 (e) 250

28 What are the common factors of 20 and 36? (2)

(a) 1, 2 and 4 (b) 1, 2, 3 and 4 (c) 1, 3 and 4 (d) 1 and 2 (e) 1, 2, and 5

29 Which of these is a cube number? (2)

(a) 25 (b) 66 (c) 125 (d) 218 (e) 121

30 Michael is collecting football stickers. There are 640 different stickers to collect altogether. He has 452 stickers, but this includes 63 doubles (i.e. he has 2 copies of 63 of the stickers). How many stickers is he missing? (2)

(a) 389 (b) 188 (c) 251 (d) 577 (e) 252

31 What is the name of this shape? (1)

(a) Cone (b) Prism (c) Sphere (d) Hemisphere (e) Cylinder

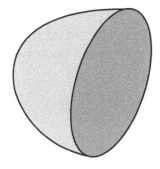

32 What is the output when 9 is put through this function machine? (2)

$9 \rightarrow -5 \rightarrow \times 6 \rightarrow ?$

(a) 4 (b) 24 (c) 64 (d) 49 (e) 180

33 $\frac{y}{3} + 7 = 12$

What is the value of $y$? (3)

(a) 12 (b) 5 (c) 15 (d) 57 (e) 9

34 Violet thought of a number, $w$. When she added 9 and then multiplied the result by 5, her final number was 60

Which expression represents her result? (2)

(a) $w \times 9 + 5 = 60$ (b) $w - 5 \times 9 = 60$ (c) $w \times 5 + 9 = 60$

(d) $5(w + 9) = 60$ (e) $5(w - 9) = 60$

35 What is the next number in this sequence? (1)

4, 16, 64, _____

(a) 80 (b) 100 (c) 256 (d) 86 (e) 106

36 Use any method to find the product of 59 and 24 (1)

(a) 35 (b) 1416 (c) 1286 (d) 1524 (e) 1466

**Turn over to the next page**

37 Use these clues to work out Penelope's favourite number. (3)

- *It is between 50 and 90*   - *It is a multiple of 8*   - *It is also a multiple of 6*

(a) 56        (b) 64        (c) 72        (d) 80        (e) 96

38 In this brick pyramid, the number on each brick is the sum of the numbers on the two bricks supporting it. What is the number on the brick marked $x$? (4)

(a) 9        (b) 12        (c) 7        (d) 5        (e) 11

| | 51 | | |
|---|---|---|---|
| 3 | 6 | 7 | X |

39 What is the remainder when 2206 is divided by 6? (1)

(a) 1        (b) 2        (c) 3        (d) 4        (e) 5

40 Three walkers are carrying water. One is carrying 330 ml, another is carrying 1500 ml and the third is carrying 850 ml. How much water do they have in total? (1)

(a) 1680 ml     (b) 2580 ml     (c) 2860 ml     (d) 2680 ml     (e) 2880 ml

41 Petra is running on a track and wants to work out her running speed. She thinks that she can calculate her speed by dividing the distance she runs by the time it takes her. What is her speed if she runs 100 metres in 25 seconds? (1)

(a) 10 m/s     (b) 5 m/s     (c) 4 m/s     (d) 2.5 m/s     (e) 2 m/s

42 In the equation $w - x + y = 30$, $w$ is three times $x$ and $y$ is one-third of $w$. What is the value of $x$? (3)

(a) 5        (b) 8        (c) 10        (d) 15        (e) 20

43 $9 \times (11 - a) = 36$

What is the value of $a$? (3)

(a) 5        (b) 6        (c) 7        (d) 8        (e) 9

44 $45 \div (j \times 3) = 3$

What is the value of $j$? (3)

(a) 7        (b) 8        (c) 5        (d) 4        (e) 15

45 What is the next number in this sequence? (1)

121                 107                 94                 82                 _____

(a) 71        (b) 80        (c) 68        (d) 69        (e) 70

46 The difference between two numbers is 37

The smaller number is 146

What is the larger number? (1)

(a) 109        (b) 173        (c) 119        (d) 183        (e) 194

47 Which multiplication in this multiplication square is incorrect? (1)

| × | 7 | 5 | 8 | 3 |
|---|---|---|---|---|
| 4 | 28 | 20 | 32 | 12 |
| 9 | 63 | 45 | 72 | 27 |
| 6 | 42 | 30 | 46 | 18 |
| 12 | 84 | 60 | 96 | 36 |

(a) 7 × 4        (b) 8 × 6        (c) 7 × 12        (d) 8 × 9        (e) 5 × 12

48 Which number between 100 and 130 is a multiple of both 8 and 7? (2)

(a) 126 (b) 128 (c) 112 (d) 102 (e) 108

49 What is $(4 + 3 \times 3^2 - 12) \times 3$? (4)

(a) 153 (b) 180 (c) 27 (d) −108 (e) 57

50 What is $18 + 6 \times 2 - 10 \div 5$? (4)

(a) 28 (b) 6.6 (c) 46 (d) 7.2 (e) 24

51 What is the next number in this sequence? (1)

144      135      126      _____

(a) 115 (b) 117 (c) 135 (d) 119 (e) 114

52 Two apples and a pear cost 80p. Three apples and a pear cost £1.06

What is the cost of a pear? (3)

(a) 26p (b) 27p (c) 28p (d) 25p (e) 30p

53 What is the next number in this sequence? (1)

0      3      8      15      _____

(a) 22 (b) 20 (c) 18 (d) 24 (e) 25

54 What is the missing number in this number sentence? (2)

$(23 \times 12) +$ _____ $= 302$

(a) 20 (b) 26 (c) 25 (d) 30 (e) 28

55 What is the missing number in this number sentence? (2)

$28 \times (23 -$ _____ $) = 532$

(a) 8 (b) 15 (c) 4 (d) 19 (e) 20

56 Work out $1026 \times 3$ (1)

(a) 3708 (b) 3078 (c) 3088 (d) 4087 (e) 4078

57 Share 7630 equally among 5 (1)

(a) 1432 (b) 1650 (c) 1426 (d) 1526 (e) 1320

58 What are the missing operations in this calculation? (2)

9 _____ 3 _____ 1 = 7 _____ 5 _____ 7

(a) ×, ×, ×, − (b) ×, −, ×, − (c) ×, +, ×, − (d) +, ×, −, × (e) ×, ×, ×, ×

59 Violet is thinking of a number, $v$. Rose is thinking of a number that is 7 less than Violet's number. Which expression shows Rose's number, $r$? (1)

(a) $r = v + 7$ (b) $r = v - 7$ (c) $v = r - 7$ (d) $v = r + 7$ (e) $r = v$

60 Ted thinks of a number. He divides it by 6 and then subtracts 4

The result is 3

What was Ted's original number? (2)

(a) 6 (b) 1 (c) 6 (d) 36 (e) 42

**Record your results and move on to the next paper**

# Paper 12

1   Write these temperatures in order, starting with the highest.   (1)

    −18°C     −23°C     −7°C     −28°C     −6°C     −11°C     −19°C

   _____   _____   _____   _____   _____   _____   _____

2   Last Tuesday, the temperature in the Sahara Desert was 52°C and the temperature in the Antarctic was −27°C. What is the difference between these temperatures?

   _____   (1)

3   $(360 \div 12) + 20 = 50$

Write down the inverse calculation you would use to check whether or not this calculation is correct? _____   (2)

4   Robbie's school is 1.36 km from his home. He walks the journey to and from school for 5 days. How far does he walk altogether? _____   (2)

5   What is $6^3 \div 6^2$? _____   (3)

6   Show that 7 is a factor of 1575 _____   (1)

7   What is the lowest common multiple of 42 and 56? _____   (2)

8   What is $\frac{5}{12} + \frac{5}{9} + \frac{3}{4}$? _____   (5)

9   What is $\frac{14}{15} - (\frac{3}{5} + \frac{1}{4})$? _____   (5)

10  What is the value of $x$ in $4^x = 256$? _____   (1)

11  $5(3x + 4) + 3(2x + 2) = 68$

What is the value of $x$? _____   (4)

12  $7(2y - 2) - 3(4y - 3) = 13$

What is the value of $y$? _____   (4)

13  Clara walked 700 m and Marcus walked 0.84 km. Write the ratio of the distance Clara walked to the distance Marcus walked in its simplest form. _____   (2)

14  Last season Jones scored 24 goals, Smith scored 28 goals and McCloud scored 20 goals. Write the ratio of goals scored by Jones : Smith : McCloud in its simplest form?

   _____   (2)

15 Reflect this pattern in the *x*-axis and then in the *y*-axis so that it appears in all four quadrants. (3)

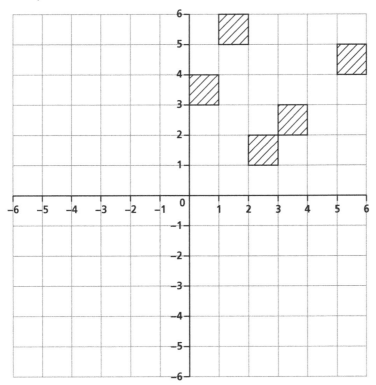

16 A circular plate has a radius of 8 cm, as shown. What is the circumference of the plate? Remember, the circumference of a circle is 3.14 × diameter. _____ (2)

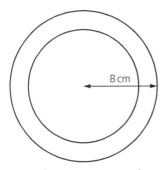

8 cm

*Not drawn accurately*

17 Donna has bought 24 cans of cola, each containing 330 ml. How many litres of cola is this? _____ (2)

18 What is the area of the 'I' shape? _____ (4)

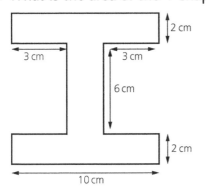

2 cm

3 cm   3 cm

6 cm

2 cm

10 cm

*Not drawn accurately*

**Turn over to the next page**

19 What is the area of this shape? _____ (3)

*Not drawn accurately*

20 An aircraft with a mass of 41 000 kg needs 14 800 kg fuel for its journey. The average mass of its 140 passengers (including luggage) is 90 kg. What is the total mass of the plane as it is about to take off? _____ (2)

21 An octagonal spinner is numbered from 1 to 8, as shown. When it is spun, what is the probability that it will land on a prime number? _____ (2)

22 Work out the value of the expression below when $a = 4$, $b = 2$ and $y = -5$.
$\sqrt{a^2 + b^3 + y^2}$ _____ (5)

23 Simplify the expression below. _____ (1)

$g \times g \times g \times h \times h \times h$

24 A cycle hire shop has bicycles (2 wheels) and tricycles (3 wheels). One Saturday morning it hires out 22 cycles with a total of 52 wheels. How many tricycles were hired? _____ (5)

25 Dusty, the hamster, eats $\frac{2}{3}$ of a bowl of hamster oats every day. How many days will it take Dusty to eat 12 bowls of hamster oats? _____ (2)

26 Henrietta receives £80 for Christmas. She buys a pair of boots costing £39.99 and a bag costing £17.49

How much money does she have left? _____ (2)

27 Which three of the numbers below have a sum of 16 and a product of 96?

_____ (2)

2   3   4   5   6   7   8   9

28 Complete this number sentence. (3)

$40 \div 8 + 4 = (107 +$ _____ $) \div 12$

29 Write the next number in this sequence? (1)

3   8   18   38   _____

**30** What is the value of angle $x$? _____ (3)

*Not drawn accurately*

**31** 4 frogs take 4 minutes to catch 4 flies. How long would it take 1 frog to catch 1 fly? _____ (1)

**32** Amari uses 3 litres of fuel when he travels 80 km on his motorbike. How much fuel will he use if he travels 120 km? _____ (2)

**33** The table shows the numbers of customers and value of sales at the Sunshine Souvenir Shop over the Easter weekend.

|  | Friday | Saturday | Sunday | Monday |
|---|---|---|---|---|
| Number of customers | 16 | 21 | 20 | 18 |
| Sales (£) | 113 | 203 | 285 | 158 |

How much does each customer spend, on average? Round your answer to the nearest 10p? _____ (4)

**34** The list below shows the number of children in each class at a secondary school.

**28    27    25    28    29    24    33    29    30    32    26    25**

What is the mean number of children in each class? _____ (2)

**35** The diameters (in kilometres) of the eight main planets in our solar system are given below.

Mercury: 4879      Venus: 12 104      Earth: 12 756      Mars: 6792

Jupiter: 142 984      Saturn: 120 536      Uranus: 51 118      Neptune: 49 528

What is the range of diameters? _____ (3)

**36** The distance between Moscow and Vladivostok on the Trans-Siberian Railway is 9300 km. The distance between Moscow and Taishet along the same route is 4500 km. Write the ratio of the distance from Moscow to Taishet : Moscow to Vladivostok in its simplest form _____ (3)

**37** A triangle has a perimeter of 78 cm. The lengths of its sides are in the ratio 3 : 4 : 5 What is the length of its longest side? _____ (3)

**38** Use the equation $v = 3.14 \times r^2 h$ to work out the volume ($v$) of the coffee jar shown. Round your answer to the nearest cm³. _____ (3)

*Not drawn accurately*

**Turn over to the next page**

39 What is the missing digit in this number statement? _____ (1)

1_2 × 15 = 2130

40 Write the missing digit in this calculation. (2)

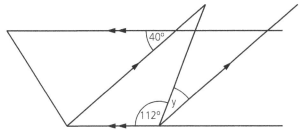

| | | 8 | 4 | 4 |
|---|---|---|---|---|
| | | 6 | 5 | 9 |
| + | | 2 | — | 8 |
| | 1 | 8 | 0 | 1 |

41 What is the value of angle $y$? _____ (3)

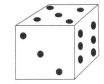

*Not drawn accurately*

42 What is the value of $v$ in the equation below? _____ (3)

$3(2v - 2) = 4(9 - 2v)$

43 Dillon rolls two normal dice and adds the scores together.

What is the probability that his total score is 6? _____ (5)

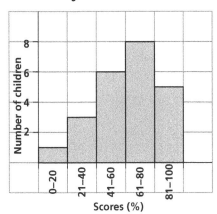

44 The results of Class 6Y's spelling test are shown in the bar chart.

How many children scored 61% or more? _____ (2)

45 Andreas thinks of a number, divides it by 8, multiplies the result by 5, subtracts 6 and
then divides this result by 2

His final number is 7

What was his original number? _____ (4)

46 Bjorn knows that he weighs 8 stone 7 pounds. How much is this in kilograms? Round your answer to the nearest kilogram. _____ (4)
   (1 stone = 14 pounds, 1 kg = 2.2 pounds)

47 Simone is 5 feet 2 inches tall. How tall is this in metres and centimetres?
   _____ (3)
   (1 inch = 2.5 cm)

48 The petrol tank in Yuri's car can hold 12 gallons of petrol. How many litres is this?
   _____ (1)
   (1 gallon = 4.5 litres)

49 This triangle has area 12 cm² and height 6 cm. How long is the base? _____ (3)

6 cm   Area of triangle = 12 cm²

*Not drawn accurately*

50 This parallelogram has an area of 45 cm². The base of the parallelogram is 7.5 cm. What is the height of the parallelogram? _____ (2)

Area = 45 cm²   h cm

7.5 cm

*Not drawn accurately*

51 A marathon race begins at 09:45 and Andy takes 3 hours, 11 minutes and 24 seconds to complete the race. Write his exact finish time in 24-hour clock format. _____ (3)

52 Complete the function machine. (2)

$4 \rightarrow \times 6 \rightarrow \boxed{\phantom{0}} \rightarrow 17$

53 Write an expression that represents the area of the rectangle. _____ (1)

y + 8 cm

y cm

*Not drawn accurately*

54 Write an expression for the shaded area? _____ (3)

y + 6

y

x − 2    x

*Not drawn accurately*

55 Simplify the expression. _____ (2)

$$\frac{6r^2 - 9r}{3r}$$

**Turn over to the next page**

56 Charlie goes out for a run that starts and finishes at his house. The graph shows his distance from home against time.

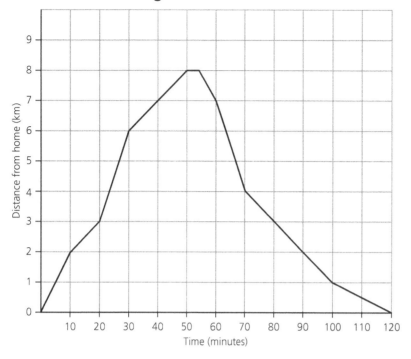

How far did Charlie run? _____ (2)

57 The two-way table shows the subjects studied by a group of students. Complete the table. (5)

|  | History | Geography | Languages | Total |
|---|---|---|---|---|
| Boys |  | 42 | 34 | 105 |
| Girls |  |  |  |  |
| Total | 76 |  | 82 | 206 |

58 The table shows the favourite milkshake flavours of a group of children. Amber wants to draw a pie chart to show the data but can't remember how to work out the sizes of the angles. Work out the size of the angle for each flavour and write it into the space provided in the table. (5)

| Ice-cream flavour | Number of children | Angle size |
|---|---|---|
| Vanilla | 12 |  |
| Strawberry | 14 |  |
| Chocolate | 16 |  |
| Cherry | 2 |  |
| Banana | 4 |  |

59 The list below shows the heights (in centimetres) of plants at a garden centre.

| 21 | 23 | 20 | 39 | 56 | 65 | 48 | 52 | 58 | 36 | 49 |
| 38 | 45 | 56 | 63 | 50 | 30 | 33 | 58 | 24 | 40 | 62 |
| 26 | 36 | 46 | 59 | 20 | 42 | 68 | 26 |

Complete the table using this data. (5)

| Plant height (cm) | Tally | Frequency |
|---|---|---|
| 20–29 | | |
| 30–39 | | |
| 40–49 | | |
| 50–59 | | |
| 60–69 | | |

60 Jess is taking part in a charity race and wants her outfit to stand out. She has a choice of a pink, purple or yellow vest and a pair of pink, purple or yellow shorts. She decides to pick a vest and a pair shorts at random. What is the probability she will pick a pink vest and pink shorts? _____ (5)

61 Write the next number in this sequence? (1)

    1    2    5    14    41    _____

62 $y = x^2 - 3x + 5$

What is the value of $y$ when $x = 5$? _____ (4)

63 $6f - 6 = 4(f + 1)$

What is the value of $f$? _____ (4)

64 The cost of staying at the Relaxing Sleep Hotel is £45 for the first night and £30 for each additional night ($n$). Write an expression to show the total cost ($t$).

_____ (1)

65 AB taxi company charges a fixed fee of £3.20 and an additional fee of 35p per mile. Thierry takes an AB taxi from the hotel to the airport and pays £9.50

How many miles is it from the hotel to the airport? _____ (2)

66 The graph shows the relationship between miles and kilometres. What is 6 miles to the nearest kilometre? _____ (1)

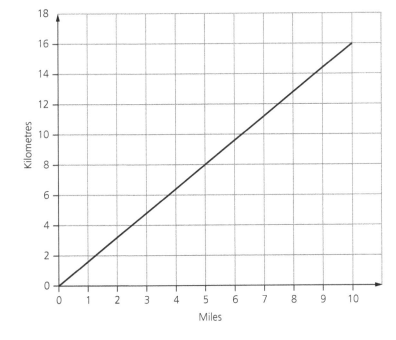

**Turn over to the next page**

**67** On the axes, draw the graph of $y = 5 - x^2$ for values of $x$ from $-3$ to $3$     (5)

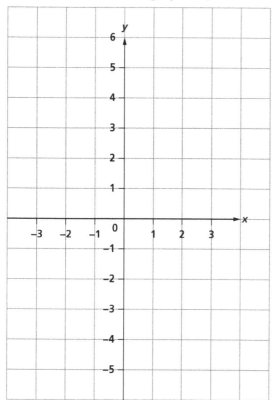

**68** Find the $n$th term for the sequence below.     (4)

| 1st | 2nd | 3rd | 4th | 5th | ... | $n$th |
|-----|-----|-----|-----|-----|-----|-------|
| 5 | 13 | 21 | 29 | 37 | ... | _____ |

**69** What is the $n$th term for this pattern? _____     (4)

Pattern

Term      1          2          3          4

**70** How many triangles will there be in the 10th pattern in this sequence?

_____     (2)

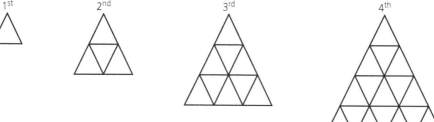

1st      2nd      3rd      4th

**Record your results and move on to the next paper**

Score [   ] / 192   Time [   ] : [   ]

# Paper 13

Test time: 75 minutes

1   The plan shows a design for a garden. Draw on all the lines of symmetry.   (4)

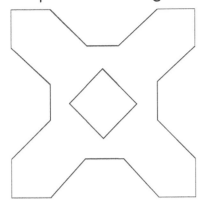

2   Find the difference between $18\frac{2}{5}$ and $35\frac{3}{5}$ _____   (1)

3   Circle the smallest number in this list.   (5)

$\frac{2}{3}$      0.333      $\frac{2}{9}$      0.24      23%      $\frac{26}{80}$      $\frac{2}{7}$

4   Add together the numbers below. Write your answer in figures (numerals).
_____   (2)

**four hundred thousand     fifty-eight thousand two hundred
eighty-seven               six million**

5   Tariq buys 5 tubs of raspberries costing £1.65 per tub and 4 packs of satsumas costing £1.89 per pack. How much change will he receive if he pays with a £20 note?
_____   (4)

6   186 plus 216 = _____   (1)

7   871 subtract 348 = _____   (1)

8   76 multiplied by 34 = _____   (1)

9   1872 divided by 12 = _____   (1)

10  Eggs are packed into boxes. Each box holds half a dozen eggs. How many boxes are needed to hold 87 eggs? _____   (2)

11  Draw hands on the clock face to show the time 20:35   (1)

12  If 3 packets of cereal cost £5.55, how much will 5 packets cost? _____   (2)

**Turn over to the next page**

**13** $N$ is a digit between 0 and 9

$$\begin{array}{r} 7\ \ N \\ \times \quad\quad N \\ \hline 4\ \ 5\ \ 6 \end{array}$$

What is the value of $N$? _____ (1)

**14** Mia asks 50 children about their favourite type of food. She draws the bar chart below to show her results. What mistake did she make when she drew the bar chart?

_____ (2)

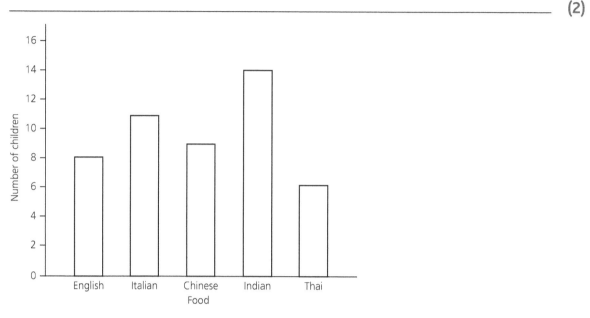

**15** Jacques wants to fill his new paddling pool. After he pours in 2271 litres of water, the pool is only $\frac{3}{8}$ full. How much water can the paddling pool hold when it is completely full? _____ (2)

**16** Ricardo uses a rope that is 32 metres long to mark out a rectangle. The length of the rectangle is three times the width. What is the area of the rectangle? _____ (3)

**17** What number is marked by the arrow on the scale below? _____ (1)

**18** Ralph has some money in his bank account. When rounded to the nearest £1000, he has £3000
What is the largest amount of money Ralph could have in his account?

_____ (1)

**19** Nina asked her friends to choose their favourite type of film. The pie chart shows the percentage of her friends that chose each type of film.

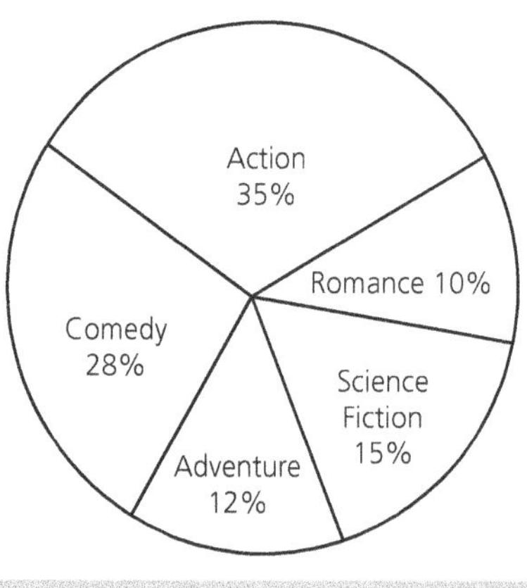

Action 35%

Romance 10%

Comedy 28%

Science Fiction 15%

Adventure 12%

If 56 of her friends chose comedy, how many friends did she ask altogether?

_____ (2)

20 What is $\frac{1}{2} + \frac{1}{3} + \frac{1}{4} + \frac{1}{8}$? Write your answer as a mixed number. _____ (5)

21 If the input for the function machine below is an odd number, will the output be even, odd or either even or odd? _____ (4)

Input → ☒ 3 → ☐+ 3 → Output

22 Farmer Phil is going to put up a post and rail fence. There will be a post every 3 metres and 3 rails between each post, as shown in the diagram. The fence will be 90 metres long. What is the combined number of posts and rails that Phil will need?

_____ (4)

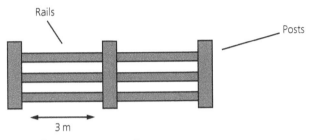

Rails

Posts

3 m

*Not drawn accurately*

23 In the magic square below, each column, row and diagonal adds up to 15
Which number will be in the cell marked with an $X$?

$X =$ _____ (3)

| | | |
|---|---|---|
| | 5 | 3 |
| 2 | | $X$ |

24 Jennifer has lots of cats and kittens. All of her cats have the same mass and all of her kittens have the same mass. Three cats and four kittens have a mass of 20 kg. Five cats and six kittens have a mass of 32 kg. What is the mass of one cat and one kitten?

_____ (5)

25 Two years ago, the combined age of five cats was 25 years. What will their combined age be in three years' time? _____ (2)

26 The number in the middle circle of each side of this triangle is the sum of the numbers on either side of that circle. Complete the diagram. (3)

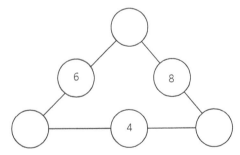

27 David and Sophie are out pushing their son, Dexter, in his pram. The wheels of the pram are 20 cm in diameter. How many turns will each wheel make as they push the pram along a 5 metre path? Round your answer to the nearest whole number.
Remember, the circumference of a circle is 3.14 multiplied by the diameter.

_____ (3)

**Turn over to the next page**

28 Through how many degrees does the minutes hand of a clock move between 4:30 pm and 5:05 pm? _____ (3)

29 Tom watches a film that lasts for 1 hour 40 minutes. Through how many degrees will the minute hand on a clock face move during this time? _____ (4)

30 Use the digits on these cards to make the closest possible number to 7000

_____ (1)

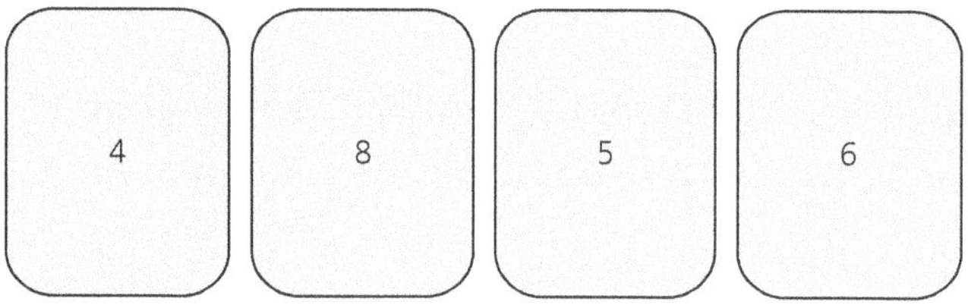

| 4 | 8 | 5 | 6 |

31 4! is an abbreviation for $4 \times 3 \times 2 \times 1$

4! = 24

Work out the value of 6! _____ (1)

32 Work out the value of $\frac{8!}{5!}$ _____ (1)

33 In the magic square below, each column, row and diagonal adds up to 34

Each of the numbers 1 to 16 can appear only once in the magic square

Complete the square and then write down the number that goes into the cell marked $X$.

$X =$ _____ (5)

| 12 |    | 14 |    |
|----|----|----|----|
|    |    |    | $X$ |
|    | 10 | 5  |    |
|    | 15 | 4  | 9  |

34 When 176 children have school dinners, regulations require that 8 dinner ladies supervise them. A new school cook starts and the number of children having school dinners increases to 286 children. How many dinner ladies will now be needed to supervise? _____ (2)

35 It takes 13 cleaners 5 days to clean the royal palace. How long will it take 10 cleaners to clean the royal palace? _____ (2)

36 There are 6 different stages to complete a 7-day race across the desert. The distance of each stage is listed below.

34 km        43.1 km        37.5 km        81.5 km        42.2 km        7.7 km

What is the mean distance of the stages? _____ (2)

37 The diagram shows a plan of Mr King's garden. He wants to install a swimming pool to replace the grass in his garden. There must be a 2-metre border around the pool. What is the greatest possible surface area of the swimming pool? _____ (5)

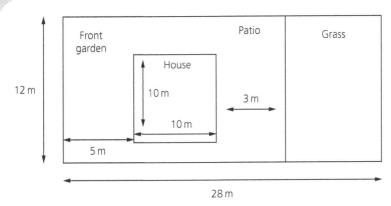

28 m

*Not drawn accurately*

**38** Translate the triangle with the vector $\begin{pmatrix} 4 \\ 3 \end{pmatrix}$. (1)

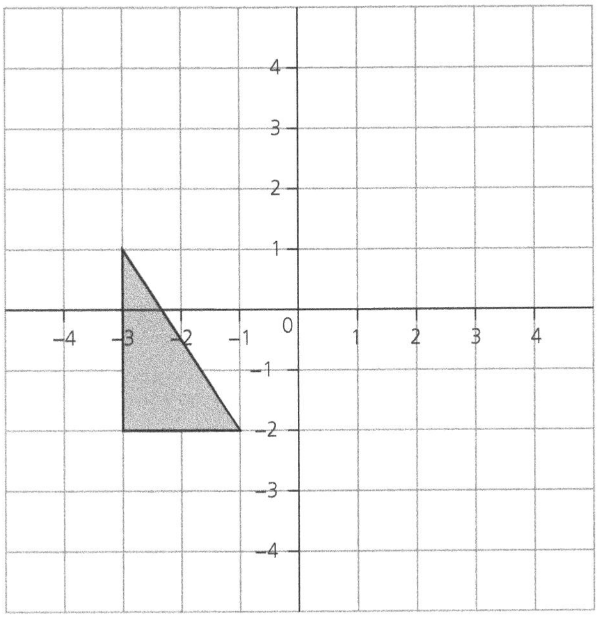

**39** $\frac{11}{13} - \frac{7}{13} =$ _____ (1)

**40** Complete the sentence below.

985 368 is 990 000, rounded to the nearest _____. (1)

**41** Jahangir is using a map to find the distance between Fromley and Gowinton. The map has a scale of 1 cm : 250 m. On the map, the distance between the two towns is 8 cm. What is the actual distance between the two towns? _____ (1)

**42** The menus from three fast food restaurants are shown below. Which restaurant is cheapest if you want to buy a cheeseburger, fries and a milkshake?

_____ (4)

| **Quick Food** | |
| --- | --- |
| Cheeseburger | £1.99 |
| Fries | £1.19 |
| Milkshake | £1.57 |

| **Speedy Bites** | |
| --- | --- |
| Cheeseburger | £1.87 |
| Fries | £1.39 |
| Milkshake | £1.43 |

| **Zoom Munch** | |
| --- | --- |
| Cheeseburger | £2.08 |
| Fries | £1.12 |
| Milkshake | £1.51 |

**Turn over to the next page**

43 The diagram shows a regular octagon that has been divided into 8 identical triangles. What is the size of angle $x$? _____ (3)

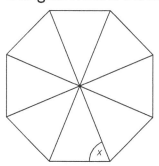

44 I toss a coin twice. What is the probability that I score two heads? _____ (2)

45 In a lucky dip at the school fair, there are 24 treat tickets, 8 trick tickets and 8 blank tickets. I pick the first ticket. What is the probability that I pick a trick ticket? Write your answer as a fraction in its simplest form. _____ (3)

46 Round 29.236 to two decimal places. _____ (1)

47 Use rounding to check whether or not the calculation below is accurate. Write down your calculation.

_____ (1)

$489 \times 9 = 44001$

48 The numbers in this sequence decrease by the same amount each time. Write down the next number in the sequence. (1)

217          157          97          37          _____

49 The Perfect Party has 44 523 supporters. They want to make themselves look as popular as possible. How could they round the number of supporters to achieve this?

_____ (1)

50 In the diagram below, Shape A has been translated to form Shape B. Write down the vector that describes the translation. $\left( \dfrac{\quad}{\quad} \right)$ (2)

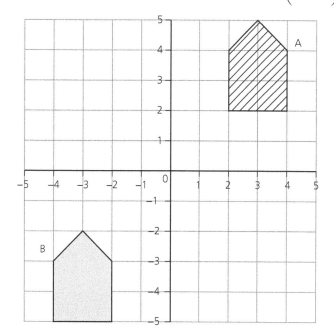

51 Aliens are landing on the Earth. There are 3-eyed aliens and 5-eyed aliens. There are 79 aliens with a total of 325 eyes. How many aliens have 3 eyes? _____ (4)

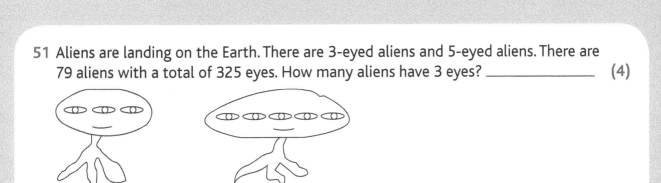

52 It takes a group of 8 children 6 days to answer all the questions in a maths book. (They work on separate pages.) How many days would you expect it to take a group of 3 children to answer all the questions? _____ (2)

53 Andy wonders how many rolls of toilet paper he would need to tear up to cover the whole area of his garden in sheets of toilet paper.

One toilet roll has 220 sheets, each measuring 12 cm by 11 cm.

Andy's garden measures 5 m by 5 m.

How many toilet rolls will Andy need? _____ (4)

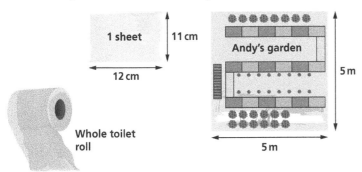

*Not drawn accurately*

54 At the pet grooming company, *Pamper your Dog*, it costs £35.00 to have a dog groomed. There is a 15% discount if a dog is groomed once a month for the whole year. How much will it cost Debra to have her dog, Roxy, groomed once a month for a whole year? _____ (3)

55 Di pays £59.40 for a dress for her brother's wedding. The dress was reduced by 50% in a sale and on the day Di bought the dress, the manager was offering a further 10% off. What was the original cost of the dress? _____ (2)

56 Hayato is 5 feet 2 inches tall and his friend Takumi is 4 feet 9 inches tall. What is the difference in their heights in centimetres? (1 inch = 2.5 cm) _____ (2)

57 The following quantities of food are recommended for 3 people.

Pasta: 375 g          Quiche: 270 g          Vegetables: $1\frac{1}{2}$ cups

What quantities of these foods would be recommended for 5 people? (3)

Pasta: _____          Quiche: _____          Vegetables: _____

**Turn over to the next page**

**58** Draw the reflection of the shape shown in the $y$-axis. (1)

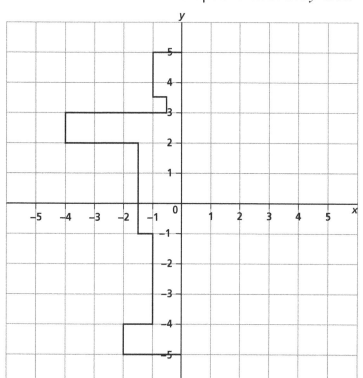

**59** Add up the amounts of money listed below. _____ (1)

    £3.48       25p       £1.75       54p       18p       £10

**60** The cost of a number of items are shown below.

    Book: £2.50       Ruler: 56p       Pen: 97p       Pad of paper: £1.84

    Map: £1.20       Bookmark: 43p       Folder: 76p

    I buy 3 different items for £3.60

    Which 3 items have I bought? _____ (1)

_____

**61** Subtract 6.78 from 23.41 _____ (1)

**62** Divide 218.4 by 6 _____ (1)

**63** Write 28 as a product of its prime factors. _____ (1)

**64** What is $\frac{9}{14} \div \frac{1}{4}$? _____ (4)

**65** Which two shapes are the same? _____ (1)

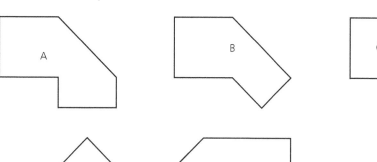

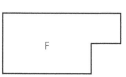

**66** These are the distances, in millions of kilometres, of the eight planets from our Sun.

| Planet | Distance from the Sun (millions km) |
|---|---|
| Mercury | 58 |
| Venus | 108 |
| Earth | 150 |
| Mars | 228 |
| Jupiter | 778 |
| Saturn | 1425 |
| Uranus | 2874 |
| Neptune | 4501 |

Which two planets are 1317 million kilometres apart?

_____ (1)

**67** DVDs cost £3.99 each. How many DVDs can be bought with £40? _____ (1)

**68** How many lines of symmetry does a regular heptagon have? _____ (1)

**69** What is the order of rotational symmetry of the shape shown? _____ (1)

**70** Duncan swims 1000 metres in a 25-metre swimming pool. He swims 15 lengths of front crawl, 10 lengths of breaststroke, 7 lengths of butterfly and the rest backstroke. How many metres of backstroke does he swim? _____ (4)

**71** Max is running a 3-kilometre race. When the winner of the race crosses the line, Max is 66% of the way through the race. How far has Max run when the winner crosses the finishing line? _____ (1)

**Turn over to the next page**

**72** The graph below shows the temperatures in a town throughout the month of April in two consecutive years. The temperatures were recorded weekly at midday. On what dates was there a difference of 2 °C between the temperatures?

_____

(1)

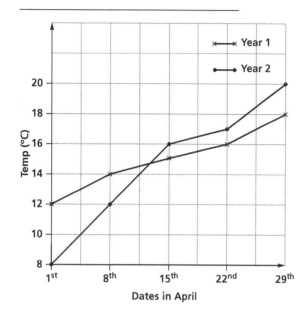

**73** Look at the grid below. Lara starts in A1 and moves across the grid following the directions. On which square does she finish? _____

(1)

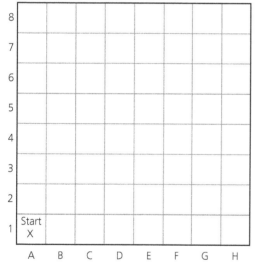

Start on A1
Move 2 squares North
Move 3 squares East
Move 2 squares North East
Move 3 squares North
Move 2 squares West
Move 1 square South
Move 2 squares South West

**74** The Carroll diagram below shows the hair length of the children in a class.

|        | Long hair | Short hair |
|--------|-----------|------------|
| Boys   | 2         | 7          |
| Girls  | 13        | 4          |

A child is picked at random. Mark with an X the position that represents the probability that the child is a girl with long hair. (2)

Impossible          Even chance          Certain

75 Krystian's office is 9 metres by 6 metres. Work out the area of his office and use this information to write down four different multiplication and division facts. (4)

_____

76 Todd emptied all the coins in his piggy bank and drew a pictogram to show the contents. How much money was in his piggy bank? _____ (5)

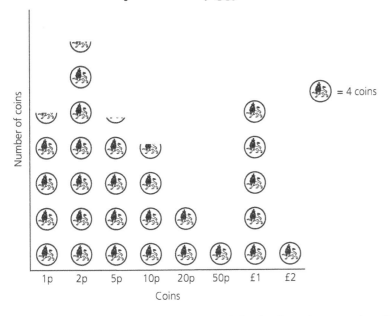

= 4 coins

Number of coins

1p    2p    5p    10p    20p    50p    £1    £2

Coins

77 Rotate the parallelogram 90° anticlockwise about point *O*. (1)

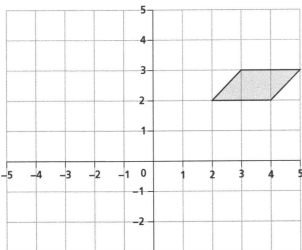

78 At a set of crossroads controlled by traffic lights, each road gets a green light in turn. When one road has a green light, the other three roads have a red light. The lights are green for 30 seconds at a time.

Jensen arrives in a queue at road 4 just as the lights turn red. Each time the traffic lights on road 4 are green, 7 cars get through the lights. Jensen is 15 cars back. How long will he have to wait to get through the lights? _____ (2)

**Turn over to the next page**

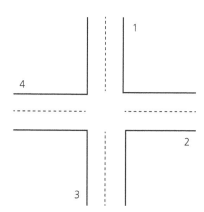

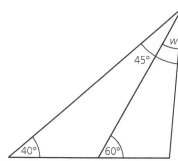

79 Work out the size of angle $w$ in this diagram. _____ (3)

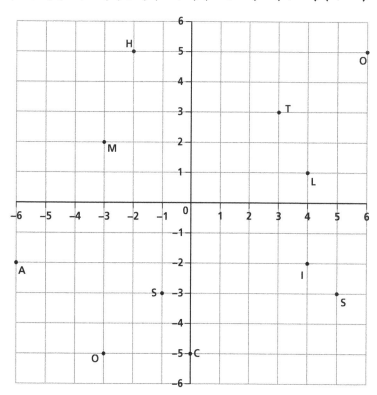

*Not drawn accurately*

80 What does the code in the grid spell out?

_____ (1)

$(-3, 2)$ $(-6, -2)$ $(3, 3)$ $(-2, 5)$ $(-1, -3)$   $(4, -2)$ $(5, -3)$   $(0, -5)$ $(6, 5)$ $(-3, -5)$ $(4, 1)$

# Answers

Guidance on the breakdown of marks is given in brackets within the questions.

Paper 10

1   17.1   7.1   7.07   1.7   0.71

2   $4\frac{3}{7}, 4\frac{1}{2}, 4\frac{2}{3}$ and $4\frac{4}{9}$ circled   $4\frac{3}{7} = 4.42$,   $4\frac{1}{2} = 4.5$,   $4\frac{3}{4} = 4.75$,   $4\frac{2}{3} = 4.66$,   $4\frac{4}{9} = 4.44$   (1) for each 2 correct

3   15 years   MMXXXV = 2035 (1)   MML = 2050 (1)   2050 – 2035 = 15 (1)

4   54 degrees   –18 °C – –72 °C

5   92p and £1.21 circled   92p (50p, 20p, 20p, 2p)   £1.21 (£1, 10p, 10p, 1p or 50p, 50p, 20p, 1p)

6   240 000 miles

7   4   Multiples of 7: 28, 84, 112 and 161

8   2 minutes 32 seconds   145 s + 18 s – 11 s = 152 s (1)   152 seconds = 2 minutes 32 seconds (1)

9   5:40   5 minutes + 20 minutes + 15 minutes + 60 minutes + 15 minutes = 115 minutes (1)
    3:45 + 115 minutes = 5:40 (1)

10  7   $4n + 7$ (1) = $5n$ (1)   $5n – 4n = 7$   $n = 7$ (1)

11  $3a + 3e + 11$

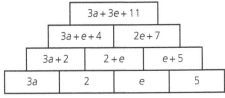

(1) for each correct row

12  500 pieces   Area of quilt: 400 cm × 300 cm = 120 000 cm² (1)   Area of small piece of material: 20 cm × 12 cm = 240 cm² (1)   120 000 cm² ÷ 240 cm² = 500 (1)

13  (1) for each 3 numbers correct

| 2 | 9 | 4 |
|---|---|---|
| 7 | 5 | 3 |
| 6 | 1 | 8 |

14  4   28 sweets ÷ (4 + 3) = 4 sweets per part (1)   4 – 3 = 1 more part, so 4 sweets (1)

15  £536   £670 × 10% = £67 (1)   20% is £67 × 2 = £134 (1)   £670 – £134 = £536 (1)

16  3 litres   47 km + 58 km + 28 km + 47 km = 180 km (1)   180 km ÷ 4 days = 45 km/day (1)
    45 km ÷ 15 km/l = 3 litres (1)

17  8A   6A (1) + ($4 \times \frac{1}{2}$ A) (1)   = 6A + 2A = 8A (1)

18  $\frac{60}{200}$ or $\frac{3}{10}$   60 + 40 + 70 + 30 = 200 (1)   Probability of a red = $\frac{60}{200} = \frac{3}{10}$ (1)

19  £5.41   3 × £1.15 = £3.45 (1)   2 × 98p = £1.96 (1)   £3.45 + £1.96 = £5.41 (1)

20

| | | |
|---|---|---|
| Not a multiple of 3 | 8 | 41, 17 |
| Multiple of 3 | 24, 48 | 27, 21, 18, 15 |
| | Multiple of 4 | Not a multiple of 4 |

21  68.24   136.48 ÷ 2

**22**

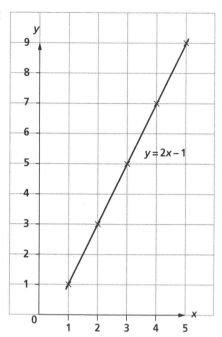

$y = 2x - 1$

**23** $\dfrac{2}{5}$   $\dfrac{48}{120}$ cancels to $\dfrac{2}{5}$

**24** certain   Wednesday will definitely follow Tuesday

**25** 365 ml

**26** 24.5 cm   2.45 cm × 10 = 24.5 cm

**27** 32 cm²   Area = base × vertical height, 8 cm × 4 cm = 32 cm²

**28** 100   0.87 × 10 = 8.7 **(1)**   870 ÷ 8.7 = 100 **(1)**

**29** 17 178   2863 × 6

**30** 1456   8736 ÷ 6

**31** $2a^2 + 10a$   $2a \times (a + 5)$ **(1)** $= 2a^2 + 10a$ **(1)**

**32**

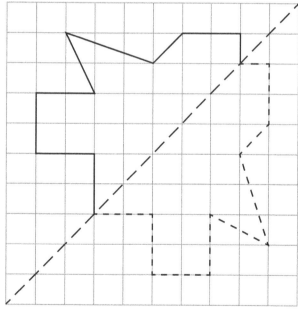

**33** 161   **(1)** for each of the criteria satisfied

**34** **(1)** for each correctly placed number

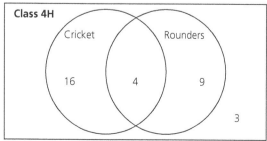

35  215   275 + 78 = 353 (1)   568 − 353 = 215 (1)

36  24 miles   1 hour 20 minutes is 4 × 20 minutes; 32 miles ÷ 4 = 8 miles (1)
    1 hour is 3 × 20 minutes, 3 × 8 miles = 24 miles (1)

37  14 °C   Highest temperature is 17 °C (1)   Lowest temperature is 3 °C (1)   17 °C − 3 °C = 14 °C (1)

38  115°   90° + 55° + 100° = 245° (1)   360° − 245° = 115° (1)

39  £3.43   180p + 95p + 68p = 343p

40  1710

| × | 30 | 8 | |
|---|---|---|---|
| 40 | 1200 | 320 | 1520 |
| 5 | 150 | 40 | 190 |
| | | | 1710 |

41  45°   105° − 60°

42  4:50 pm   7:30 pm − 2 hours = 5:30 pm (1)   5:30 pm − 40 minutes = 4:50 pm (1)

43  (3, 4)   (2, 1)   (6, 5)   (5, 2)   (1, 2)   (1) for each pair of co-ordinates in correct position

44  $\frac{1}{2}$   3 (2, 4 and 6) out of 6 numbers so $\frac{3}{6}$ (1)   $= \frac{1}{2}$ (1)

45  5026 m   3.2 km = 3200 m (1)   3200 m + 1826 m = 5026 m (1)

46  37 children   7 + 12 + 5 + 6 + 7 (1)   = 37 (1)

47  $F = 8$   $6F − 42 = 6$ (1)   $6F = 48$ (1)   $F = 8$ (1)

48  $W = −1$   $8W + 12 = 4W + 8$ (1)   $8W − 4W = 8 − 12$ (1)   $4W = −4$ (1)   $W = −1$ (1)

49  67.5   49 + 35 + 109 + 78 + 38 + 0 + 145 + 5 + 189 + 27 = 675 (1)   675 ÷ 10 = 67.5 (1)

50  1200–1300   The steepest part of the graph

## Paper 11

1   (d) 88 237 this number is rounded down to 88 000

2   (d) 25   10 km = 10 000 m (1)   10 000 m ÷ 400 m = 25 (1)

3   (d) 361   3 flashes a minute (1) × 120 minutes = 360 flashes (1)   one flash at the beginning, so 361 (1)

4   (e) 24   $p + c = 32$, where $p$ is the number of pigs and $c$ is the number of chickens (1)   $4p + 2c = 80$ (1)
    Solve by trial and error or rearrange first equation ($c = 32 − p$ (1))   and solve the second by substitution
    ($4p + 2(32 − p) = 80$, $4p + 64 − 2p = 80$, $2p = 16$, $p = 8$ (1)   so $c = 32 − 8 = 24$ (1)

5   (e) Kite because it doesn't have any parallel sides

6   (c) Tuesday will follow Monday.

7   (b) 69   92 ÷ 4 = 23, 23 × 3 = 69

8   (b) 2848

9   (b) 86

10  (d) $y = x − 3$   (1) for at least 2 pairs of co-ordinate values

| $x$ | −2 | −1 | 0 | 1 | 2 | 3 | 4 | 5 |
|---|---|---|---|---|---|---|---|---|
| $y$ | −5 | −4 | −3 | −2 | −1 | 0 | 1 | 2 |

11  (a) 15 boys   8 + 3 = 11 parts (1)   55 children ÷ 11 = 5 (1)   5 × 3 = 15 (1)

12  (a) 10 litres   15 + 1 = 16 parts (1)   160 litres ÷ 16 = 10 litres (1)

13  (d) 68%   $\frac{170}{250} = \frac{17}{25}$ (1)   $\frac{17}{25} = \frac{68}{100}$ = 68% (1)

14  (e) 26.78

15  (a) 144 cm   Side length of square is 18 cm, so side length of equilateral triangle is 18 ÷ 3 = 6 cm (1)
    Perimeter of this shape is 6 cm × 24 = 144 cm (1)

16  (b) 360 pages   135 ÷ 3 = 45 (1)   45 × 8 = 360 (1)

17  (a) 19.6 m   28 × 70 = 1960 cm (1)   1960 cm ÷ 100 = 19.6 m (1)

18  (b) 84   4 and 7 are both factors 28 exactly (4 × 7 = 28) (1)   84 is a multiple of 28 between 80 and 90 (1)

19 (c) 54°   Corresponding angles are equal (1)   Angles on a straight line are 180°, 180° − 126° = 54° (1)

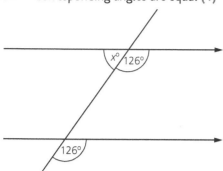

20 (d) 46°   Corresponding angles are equal (1)   Angles in a triangle add up to 180°, 180° − (95° + 39°) = 46° (1)

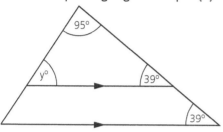

21 (b) £621   $900 ÷ 1.45 = £620.68 (1)   Rounded to the nearest £ = £621 (1)

22 (d) $\frac{14}{15}$   $\frac{18}{30} + \frac{10}{30}$ (1)   $= \frac{28}{30}$ (1) $= \frac{14}{15}$ (1)

23 (b) $1\frac{7}{20}$   $1\frac{15}{20} - \frac{8}{20}$ (1) $= 1\frac{7}{20}$ (1)

24 (a) 16   403 parents ÷ 26 seats = 15.5 (1)   16 rows needed (1)
25 (b) 11.25 miles   18 km ÷ 1.6 km
26 (a) £474   158 × £3
27 (c) 225   15 × 15
28 (a) 1, 2 and 4   Factors of 20: 1, 2, 4, 5, 10 & 20 (1)   Factors of 36: 1, 2, 3, 4, 6, 9, 12, 18 & 36 (1)
29 (c) 125   (1) for evidence of working out 3³ or 4³
30 (c) 251   452 − 63 = 389 (1)   640 − 389 = 251 (1)
31 (d) Hemisphere
32 (b) 24   9 − 5 = 4 (1)   4 × 6 = 24 (1)
33 (c) 15   $\frac{y}{3}$ = 12 − 7 = 5 (1)   y = 5 × 3 (1)   = 15 (1)
34 (d) 5(w + 9) = 60   (1) for (w + 9) seen
35 (c) 256   Each previous number is multiplied by 4, so 64 × 4
36 (b) 1416
37 (c) 72   (1) for each of the criteria satisfied
38 (a) 9

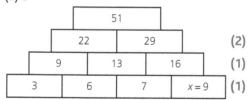

39 (d) 4   367 remainder 4
40 (d) 2680 ml
41 (c) 4 m/s   Speed = Distance (100 m) ÷ Time (25 s)
42 (c) 10   w = 3x (1)   y = 3x ÷ 3 = x (1)   3x − x + x = 30, x = 10 (1)
43 (c) 7   99 − 9a = 36 (1)   9a = 63 (1)   a = 7 (1)
44 (c) 5   45 ÷ 3j = 3 (1)   45 ÷ 15 = 3 (1)   and   5 × 3 = 15 so j = 5 (1)
45 (a) 71   The difference between the terms decreases by 1 each time
46 (d) 183
47 (b) 8 × 6   8 × 6 = 48
48 (c) 112   112 = 8 × 14 (1) and 7 × 16 (1)
49 (e) 57   (4 + 3 × 9 − 12) × 3 (1)   (4 + 27 − 12) × 3 (1)   19 × 3 (1)   = 57 (1)
50 (a) 28   18 + 6 × 2 − 2 (1)   18 + 12 − 2 (1)   30 − 2 (1)   = 28 (1)

51  (b) 117  Each number decreases by 9
52  (c) 28p  Cost of apple: £1.06 – 80p = 26p (1)  Cost of pear is 80p – (2 × 26p) (1)  = 80p – 52p = 28p (1)
53  (d) 24  Differences each time increases by 2
54  (b) 26  23 × 12 = 276 (1)  302 – 276 = 26 (1)
55  (c) 4  532 ÷ 28 = 19 (1)  23 – 19 = 4 (1)
56  (b) 3078
57  (d) 1526
58  (c) ×, +, ×, –  9 × 3 + 1 = 28 (1)  7 × 5 – 7 = 28 (1)
59  (b) $r = v – 7$
60  (e) 42  3 + 4 = 7 (1)  7 × 6 = 42 (1)

## Paper 12

1  –6°C  –7°C  –11°C  –18°C  –19°C  –23°C  –28°C
2  79°C  52°C – –27°C = 52°C + 27°C = 79°C
3  (50 – 20) × 12 = 360 (1) for each operation
4  13.6 km  5 × 2 = 10 journeys (1)  1.36 km × 10 = 13.6 km (1)
5  6  $6^3 ÷ 6^2 = (6 × 6 × 6)$ (1) ÷ (6 × 6) (1)  = 216 ÷ 36 = 6 (1)
6  1575 ÷ 7 = 225  whole number answer
7  168  42, 84, 126, **168** (1)  56, 112, **168** (1)
8  $1\frac{13}{18}$  $\frac{15}{36} + \frac{20}{36} + \frac{27}{36}$ (2) $= \frac{62}{36}$ (1) $= 1\frac{26}{36}$ (1) $= 1\frac{13}{18}$ (1)
9  $\frac{1}{12}$  $\frac{14}{15} - \left(\frac{12}{20} + \frac{5}{20}\right)$ (1) $= \frac{14}{15} - \frac{17}{20}$ (1) $= \frac{56}{60} - \frac{51}{60}$ (1) $= \frac{5}{60}$ (1) $= \frac{1}{12}$ (1)
10  4  4 × 4 × 4 × 4 = 256
11  2  $15x + 20 + 6x + 6 = 68$ (2)  $21x = 68 – 26 = 42$ (1)  $x = 2$ (1)
12  9  $14y – 14 – (12y – 9) = 13$ (1)  $14y – 14 – 12y + 9 = 13$ (1)  $2y = 13 + 5 = 18$ (1)  $y = 9$ (1)
13  5:6  700:840 (1)  = 5:6 (1)
14  6:7:5  24:28:20 (1)  simplifies to 6:7:5 (1)
15

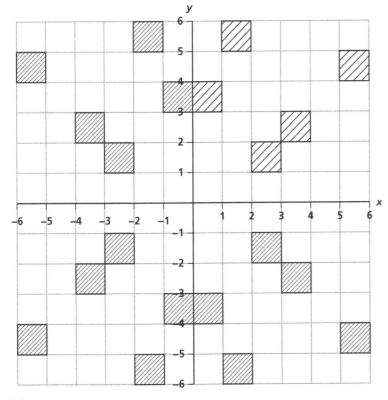

(1) for each correct quadrant
16  50.24 cm  diameter = 2 × radius = 2 × 8 cm = 16 cm (1)
circumference = 3.14 × diameter = 3.14 × 16 cm = 50.24 cm (1)
17  7.92 litres  24 × 330 ml = 7920 ml (1)  7920 ml ÷ 1000 = 7.92 litres (1)
18  64 cm²  top / bottom part: 10 cm × 2 cm = 20 cm² (1)  20 cm² × 2 = 40 cm² (1)
middle part: 6 cm × 4 cm = 24 cm² (1)  total area: 40 cm² + 24 cm² = 64 cm² (1)

19  54 cm²   area of rectangle: 9 cm × 4 cm = 36 cm² (1)   area of triangle = $\frac{1}{2}$ × base × vertical height = $\frac{1}{2}$ × 9 cm × 4 cm = $\frac{1}{2}$ × 36 cm² = 18 cm² (1)   Total area: 36 cm² + 18 cm² = 54 cm² (1)

20  68 400 kg   Mass of people and luggage: 140 × 90 kg = 12 600 kg (1)
Total mass = 41 000 kg + 14 800 kg + 12 600 kg = 68 400 kg (1)

21  $\frac{4}{8}$ or $\frac{1}{2}$   4 prime numbers: 2, 3, 5 and 7 (1)   4 chances in 8 (1)

22  7   $\sqrt{16}$(1) + 8(1) + 25(1) = $\sqrt{49}$ (1) = 7 (1)

23  $g^3h^3$

24  8   $t + b = 22$, where $t$ stands for tricycles and $b$ for bicycles (1)   $3t + 2b = 52$ (1)   Rearrange one equation $b = 22 - t$ (1)   and substitute into the other to solve: $3t + 2(22 - t) = 52$ (1)   $3t + 44 - 2t = 52$, $t = 52 - 44 = 8$ (1)

25  18 days   12 ÷ $\frac{2}{3}$ = 12 × $\frac{3}{2}$ (1)   = 36 ÷ 2 = 18 (1)

26  £22.52   £39.99 + £17.49 = £57.48 (1)   £80 − £57.48 = £22.52 (1)

27  2, 6 and 8   2 + 6 + 8 = 16 (1)   2 × 6 × 8 = 96 (1)

28  1   5 + 4 = (107 + ___ ) ÷ 12 (1)   9 = (107 + ___ ) ÷ 12 (1)   9 × 12 = 108, 108 − 107 = 1 (1)

29  78   look for a relationship between the differences or multiply previous term by 2 and + 2

30  74°   62° + 88°+ 104° = 254° (1)   360° in a quadrilateral, so 360° − 254° = 106° (1)   180° on a straight line, so 180° − 106° = 74° (1)

31  4 minutes   Each frog takes 4 minutes to catch a single fly

32  $4\frac{1}{2}$ litres   80 km uses 3 litres of fuel, so 40 km uses 3 ÷ 2 = $1\frac{1}{2}$ litres (1) and 120 km uses 3 × $1\frac{1}{2}$ litres (1)

33  £10.10   £759 (1) ÷ 75 (1) = £10.12 (1) = £10.10, to the nearest 10p (1)

34  28 children   336 (1) ÷ 12 = 28 (1)

35  138 105 km   142 984 km (1) − 4879 km (1) = 138 105 km (1)

36  15:31   4500 km : 9300 km (1) = 45:93 (1)   = 15:31 (1)

37  32.5 cm   3 + 4 + 5 = 12 parts (1)   78 cm ÷ 12 = 6.5 cm (1)   6.5 cm × 5 = 32.5 cm (1)

38  653 cm³   $v$ = 3.14 × (4 cm)² × 13 cm (1)   $v$ = 653.12 cm³ (1)   = 653 cm³ (1)

39  4   2130 ÷ 15 = 142

40  9   844 + 659 = 1503 (1)   1801 − 1503 = 298 (1)

41  28°   Angle $a$ = 180° − 40° = 140° (angles on a straight line) (1)   angle $a$ = angle $w$ (parallel lines) (1)
angle $y$ = 140° − 112° = 28° (1) *or alternative method*

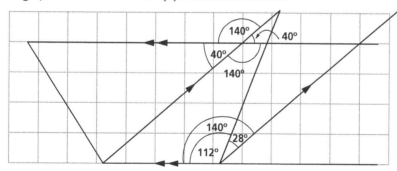

42  3   $6v − 6 = 36 − 8v$ (1)   $14v = 42$ (1)  $v = 3$ (1)

43  $\frac{5}{36}$   (1) for each of the 5 ways to score 6 seen or total of 36 combinations

44  13   8 + 5 (1)   = 13 (1)

45  32   Use the inverse operations: So 7 ⏹×⏹ 2 (1) = 14 ⏹+⏹ 6 (1) = 20 ⏹÷⏹ 5 (1) = 4 ⏹×⏹ 8 (1) = 32

46  54 kg   8 stone 7 lb = 8 × 14 lb + 7 lb = 112 lb (1) + 7 lb = 119 lb (1)   119 lb ÷ 2.2 = 54.09 kg (1) = 54 kg (1)

47  1 m 55 cm   5 feet 2 inches = 5 × 12 + 2 = 60 inches (1) + 2 inches = 62 inches (1)   62 × 2.5 = 155 cm (1)

48  54 litres   12 × 4.5

49  4 cm   $A = \frac{1}{2}$ × $b$ × $h$, so 12 cm² = $\frac{1}{2}$ × $b$ × 6 cm (1)   $3b$ = 12 cm (1)   $b$ = 4 cm (1)

50  6 cm   $A = b$ × $h$, so 45 cm² = 7.5 cm × $h$ (1)   $h$ = 6 cm (1)

51  12:56:24   09:45 + 3 hours = 12:45 (1)  12:45 + 11 minutes = 12:56 (1)  12:56 + 24 seconds = 12:56:24 (1)

52  ⏹− 7⏹   4 × 6 = 24 (1)   24 − 7 = 17 (1)

53  $y(y + 8)$ *or* $y^2 + 8y$

54  $6x + 2y$   Area of larger rectangle: $x(y + 6)$ *or* $xy + 6x$ (1)   Area of smaller rectangle: $y(x − 2)$ *or* $xy − 2y$ (1)
Shaded area: $(xy + 6x) − (xy − 2y) = 6x + 2y$ (1)

55  $2r − 3$   $6r^2 ÷ 3r = 2r$ (1)   $9r ÷ 3r = 3$ (1)

56 16km   8km (1) from home   8km back home (1)

57 (2) for both History values correct, (2) for both Geography values correct, (1) for missing Languages and Total correct

|  | History | Geography | Languages | Total |
|---|---|---|---|---|
| Boys | **29** | 42 | 34 | 105 |
| Girls | **47** | **6** | **48** | **101** |
| Total | 76 | **48** | 82 | 206 |

58 If no answers correct, (1) for total of 48 children and (1) for 7.5°/child

| Ice-cream flavour | Number of children | Angle size | |
|---|---|---|---|
| Vanilla | 12 | 360° × 12 ÷ 48 = 90° | (1) |
| Strawberry | 14 | 360° × 14 ÷ 48 = 105° | (1) |
| Chocolate | 16 | 360° × 16 ÷ 48 = 120° | (1) |
| Cherry | 2 | 360° × 2 ÷ 48 = 15° | (1) |
| Banana | 4 | 360° × 4 ÷ 48 = 30° | (1) |

59

| Plant height (cm) | Tally | Frequency | |
|---|---|---|---|
| 20–29 | ШΙΙ | 7 | (1) |
| 30–39 | ШΙ | 6 | (1) |
| 40–49 | ШΙ | 6 | (1) |
| 50–59 | ШΙΙ | 7 | (1) |
| 60–69 | ΙΙΙΙ | 4 | (1) |

60 $\frac{1}{9}$ (pink vest , pink shorts),(pink vest, purple shorts), (pink vest, yellow shorts) (1)   (purple vest, pink shorts), (purple vest, purple shorts), (purple vest, yellow shorts) (1)   (yellow vest, pink shorts)   (yellow vest, purple shorts)   (yellow vest, yellow shorts) (1), 9 different combinations (1), 1 is pink vest, pink shorts (1)

61 122   Multiply the previous term by 3 and subtract 1 or work out and continue the pattern of differences (+ 1, + 3, + 9, + 27; the difference is 3 times bigger each time)

62 15 $y = 5^2 - (3 \times 5) + 5$ (1) $y = 25 - (15 + 5)$ Applying BIDMAS (1) $y = 25 - 20$ (1) $y = 5$ (1) $= 10 + 5$ (1) $= 15$ (1)

63 5   $6f - 6$ (1)   $= 4f + 4$ (1)   $2f = 10$ (1)   $f = 5$ (1)

64 $t = 45 + 30n$

65 18 miles   £9.50 – £3.20 = £6.30 (1)   630p ÷ 35p = 18 (1)

66 10km

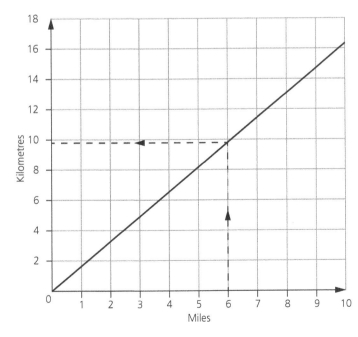

67  (2) for all co-ordinates correct, or (1) for at least 4 correct    (2) for plotting all co-ordinates correctly or (1) for at least 4 co-ordinates plotted correctly, (1) curve correctly drawn

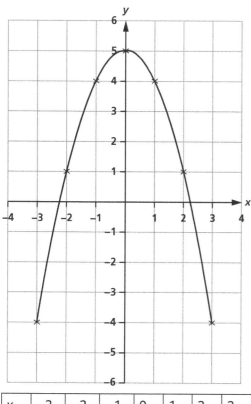

| x | −3 | −2 | −1 | 0 | 1 | 2 | 3 |
|---|----|----|----|---|---|---|---|
| y | −4 | 1  | 4  | 5 | 4 | 1 | −4 |

68  $8n − 3$   Difference between each term is 8 (1)   so $n$th term based on $8n$ (1)   When $n = 1$, $8n = 8$ and first term is 5 which is 3 less than 8 (1)   $n$th term is $8n − 3$ (1)

69  $3n + 1$   The number of matchsticks in each diagram increases by 3 (1)   so $n$th term based on $3n$ (1)   When $n = 1$, $3n = 3$ and first term is 4 which is 1 more than 3 (1)   $n$th term is $3n + 1$ (1)

70  100   each term squared gives the number of triangles (1), so $n$th term is $n^2$ (1)


Paper 13

1

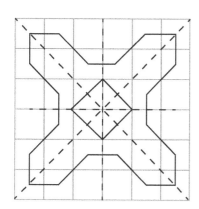

(1) for each of four correct lines of symmetry

2   $17\frac{1}{5}$

3   $\frac{2}{9}$ $\frac{2}{3} = 0.666$ (1)   $0.333$   $\frac{2}{9} = 0.222$ (1)   $0.24$   $23\% = 0.23$   $\frac{26}{80} = 0.325$ (1)   $\frac{2}{7} = 0.285$ (1)   so $\frac{2}{9}$ (1)

4   6 458 287   400 000 + 58 200 + 87 + 6 000 000 (1)   = 6 458 287 (1)

5  £4.19    £1.65 × 5 = £8.25 (1)    £1.89 × 4 = £7.56 (1)    Total spent: £8.25 + £7.56 = £15.81 (1)
   Change: £20.00 − £15.81 = £4.19 (1)
6  402    186 + 216
7  523    871 − 348
8  2584    76 × 34
9  156    1872 ÷ 12
10  15 boxes    87 ÷ 6 = 14 remainder 3 (1)    15 boxes needed (1)
11

12  £9.25    1 packet costs £5.55 ÷ 3 = £1.85 (1)    £1.85 × 5 = £9.25 (1)
13  6    76 × 6 = 456
14  The bar chart shows only 48 children not 50 (1)    8 + 11 + 9 + 14 + 6 = 48 (1)
15  6056 litres    2271 litres ÷ 3 = 757 (1)    757 × 8 = 6056 litres (1)
16  48 m²    length + width = 32 m ÷ 2 = 16 m, 16 m ÷ 4 = 4 m (1)    so width is 4 m and length is 4 m × 3 = 12 m (1)
   area = 4 m × 12 m = 48 m² (1)
17  −35
18  £3499.99    £3500 round up to £4000
19  200    56 ÷ 28% (1) × 100% = 200 (1)
20  $1\frac{5}{24}$    $\frac{12}{24}$ (1) + $\frac{8}{24}$ (1) + $\frac{6}{24}$ (1) + $\frac{3}{24}$ (1) = $\frac{29}{24}$ = $1\frac{5}{24}$ (1)
21  even    (1) for each correct example, up to 3 marks
22  121 posts & rails    31 posts (2) [award (1) for 30 posts]    30 gaps, so 30 × 3 (1) = 90 rails    31 + 90 = 121 (1)
23  $X = 4$    (2) for 8 in top right corner

| 6 | 1 | 8 |
|---|---|---|
| 7 | 5 | 3 |
| 2 | 9 | 4 |

24  6 kg    $3c + 4k = 20$ (1) and $5c + 6k = 32$ (1)
   Rearrange: $c = \frac{20 - 4k}{3}$ (1)
   Substitute and solve to find $k$: $5 \left(\frac{20 - 4k}{3}\right) + 6k = 32$ (1)    so $\frac{100 - 20k}{3} + 6k = 32$    $100 - 20k + 18k = 96$
   $2k = 4$ $k = 2$
   Substitute to find $c$:    $c = \frac{20 - 4k}{3} = \frac{20 - 8}{3} = \frac{12}{3} = 4$
   $c + k = 6$ (1)
25  50 years    5 years elapsed for 5 cats = 25 years (1)    25 years + 25 years (1)
26  5 (1)    3 (1)    1 (1)

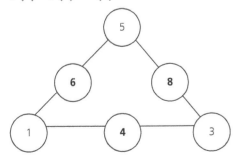

27  8 turns    Distance travelled by wheel in 1 turn is the circumference: 3.14 × 20 cm = 62.8 cm (1)
   Number of turns in 5 m (500 cm) is 500 ÷ 62.8 cm = 7.96 turns (1)    8 turns (1)
28  210°    Minute hand turns 90° in 15 minutes (1)    90 ÷ 3 = 30° in 5 minutes (1)    90° + 90° + 30° = 210°
29  600°    360° in 1 hour (1)    180° in 30 minutes (1)    60° in 10 minutes (1)    360° + 180° + 60° = 600° (1)
30  6854
31  720    6 × 5 × 4 × 3 × 2 × 1

**32** 336 $8 \times 7 \times 6$

**33** $X = 2$ (1) for each two correctly placed numbers if $X$ is incorrect

| 12 | 1 | 14 | 7 |
|----|----|----|----|
| 13 | 8 | 11 | 2 |
| 3 | 10 | 5 | 16 |
| 6 | 15 | 4 | 9 |

**34** 13    176 children ÷ 8 dinner ladies = 22 children per dinner lady (1)    286 children ÷ 22 children per dinner lady = 13 dinner ladies required (1)

**35** 6.5 days    13 cleaners × 5 days = 65 working days (1)    65 days ÷ 10 cleaners = 6.5 days (1)

**36** 41 km    34 km + 43.1 km + 37.5 km + 81.5 km + 42.2 km + 7.7 km = 246 km (1)    246 km ÷ 6 = 41 km (1)

**37** 48 m²    5 m + 10 m + 3 m = 18 m (1)    Available area is 10 m × 12 m (1)    2 m border around the pool gives dimensions 6 m (1) × 8 m (1) = 48 m² (1)

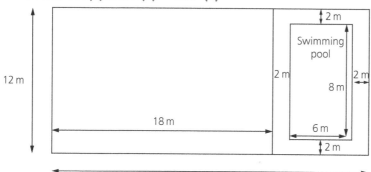

**38**

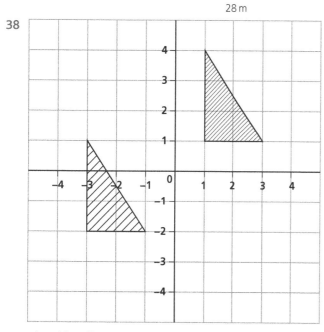

**39** $\frac{4}{13}$    $\frac{11}{13} - \frac{7}{13}$

**40** 10 000

**41** 2000 m or 2 km    $8 \times 250$ m

**42** Speedy Bites    Quick Food: £4.75 (1)    Speedy Bites: £4.69 (1)    Zoom Munch: £4.71 (1)    Speedy Bites is cheaper (1)

43  67.5°  Angles at centre: 360° ÷ 8 = 45° (1)   All internal triangles are isosceles, so 180° − 45° = 135°(1)
135° ÷ 2 = 67.5° (1) *Alternative methods are possible*

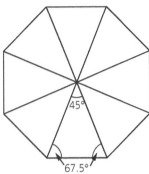

44  $\frac{1}{4}$  Possible outcomes: HH, TT HT, TH (1)   $\frac{1}{4}$ (1)

45  $\frac{1}{5}$  24 + 8 + 8 tickets = 40 tickets (1)   $\frac{8}{40}$ (1)   $= \frac{1}{5}$ (1)

46  29.24

47  Not accurate, 500 × 10 = 5000

48  −23

49  Round to nearest 1000, 45 000 supporters

50  $\begin{pmatrix} -6 \\ -7 \end{pmatrix}$  (1) for each correct number

51  35  $T + F = 79$, where $T$ is a 3-eyed alien and $F$ is a 5-eyed alien (1)   $3T + 5F = 325$ (1)   $F = 79 − T \rightarrow 3T + 5(79 − T)$
= 325 (1) → 3$T$ + 395 − 5$T$ = 325 → 2$T$ = 70, $T$ = 35 (1)

52  16 days   8 × 6 = 48 days (1)   48 ÷ 3 = 16 days(1)

53  9 rolls   Area of one sheet of toilet paper: 12 cm × 11 cm = 132 cm² (1)   Area of one roll: 132 cm² × 220 =
29 040 cm² = 2.9 m² (1)   Area of garden: 5 m × 5 m = 25 m² (1)
Number of rolls needed: 25 m² ÷ 2.9 m² = 8.6 → 9 rolls (1)

54  £357   £35 × 12 = £420 (1)   15% of £420 = £63 (1)   Cost for year: £420 − £63 = £357 (1)

55  £132   Cost: £59.40 ÷ 90 × 100 = £66 (1)   £66 × 2 (reduced by 50%) = £132 (1)

56  12.5 cm   5 feet 2 inches − 4 feet 9 inches = 5 inches (1)   5 inches × 2.5 = 12.5 cm (1)

57  Pasta: 375 g ÷ 3 = 125 g × 5 = 625 g (1)   Quiche: 270 g ÷ 3 = 90 g × 5 = 450 g (1)
Vegetables: $1\frac{1}{2}$ cups ÷ 3 = $\frac{1}{2}$ cup × 5 = $2\frac{1}{2}$ cups (1)

58

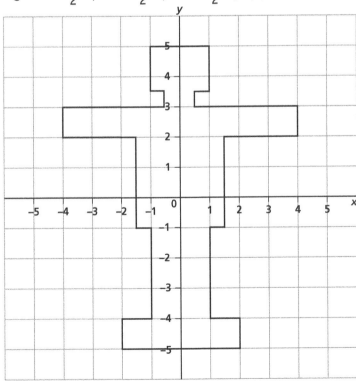

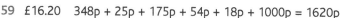

59  £16.20   348p + 25p + 175p + 54p + 18p + 1000p = 1620p
60  Ruler, Map, Pad of paper   56p + £1.20 + £1.84 = £3.60
61  16.63
62  36.4
63  $2 \times 2 \times 7$
64  $2\frac{4}{7}$   $\frac{9}{14} \times \frac{4}{1}$ (1)   $= \frac{36}{14}$ (1)   $= 2\frac{8}{14}$ (1)   $= 2\frac{4}{7}$ (1)
65  B and D

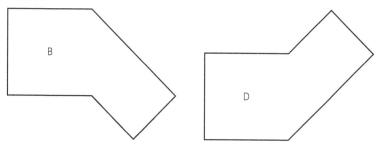

66  Saturn and Venus   1425 million km − 108 million km = 1317 million km
67  10 DVDs   £3.99 × 10 = £39.99
68  7

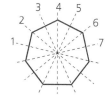

69  4   When turned the shape will look the same on 4 occasions
70  200 m   1000 ÷ 25 = 40 lengths (1)   15 + 10 + 7 = 32 lengths swum (1)   40 − 32 = 8 lengths (1)
     8 × 25 = 200 m of backstroke (1)
71  1.98 km   $3 \text{ km} \times \frac{66}{100} = 1.98 \text{ km}$
72  8th April, 29th April
73  B5

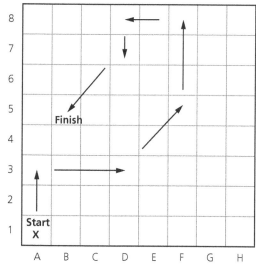

74

     13 out of 26 girls have long hair $\frac{1}{2}$ (1)   Correctly marked on scale (1)
75  9 × 6 = 54 (1)   6 × 9 = 54 (1)   54 ÷ 9 = 6 (1)   54 ÷ 6 = 9 (1)

**76** £34.65   (1p × 18) + (2p × 26) + (5p × 17) + (10p × 15) + (20p × 8) + (50p × 4) + (£1 × 20) + (£2 × 4) = 18p + 52p + 85p + £1.50 + £1.60 + £2.00 + £20.00 + £8.00 = £34.65 **(5)** for the correct final total, 0 if incorrect   **(1)** for correct totals of two coin types, **(2)** for correct totals of four coin types, **(3)** for correct totals of six coin types, **(4)** for correct totals of eight coin types

**77**

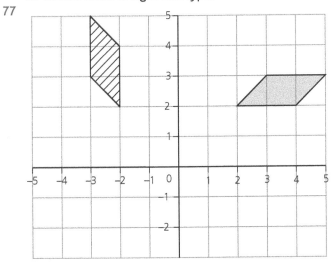

**78** 5 minutes 30 seconds   Jensen will get through on his third green light **(1)**   2 minutes elapse for all 4 roads to have a green light, 2 minutes × 2 = 4 minutes + 1 minute 30 seconds for third green light **(1)**

**79** 25°   Angle $A$ = 180° − 60° = 120° **(1)**   Angle $B$ = 180° − (120° + 40°) = 20° **(1)**   Angle $w$ = 45° − 20° = 25° **(1)**

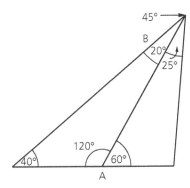

**80** MATHS IS COOL